THE QUEST FOR CORTISONE

THE QUEST FOR CORTISONE

Thom Rooke

Michigan State University Press
East Lansing

⊗ The paper used in this publication meets the minimum requirements of ANSI/NISO Z39.48-1992 (R 1997) (Permanence of Paper).

 Michigan State University Press
East Lansing, Michigan 48823-5245

Printed and bound in the United States of America.

18 17 16 15 14 13 12   1 2 3 4 5 6 7 8 9 10

LIBRARY OF CONGRESS CATALOGING-IN-PUBLICATION DATA
Rooke, Thom W.
The quest for cortisone / Thom Rooke.
    p. cm.
Includes bibliographical references.
    ISBN 978-1-61186-033-7 (pbk. : alk. paper)
    I. Title.
    1. Kendall, Edward C. (Edward Calvin), 1886-1972. 2. Hench, Philip S. (Philip Showalter), 1896-1965. 3. Mayo Clinic. 4. Cortisone—history—United States. 5. Clinical Trials as Topic—history—United States. 6. Cortisone—therapeutic use—United States. 7. Drug Discovery—history—United States. 8. Drug Evaluation—history—United States. 9. History, 20th Century—United States.
    610.72'4—dc23   2011031277

Book design by Scribe Inc (www.scribenet.com)
Cover design by David Drummond, Salamander Design, Inc.
Cover photo of Edward C. Kendall, PhD, and Philip S. Hench, MD, in Dr. Kendall's laboratory, taken November 15, 1949, after the announcement of their receipt of the 1950 Nobel Prize in Physiology or Medicine for the discovery of cortisone. Used with permission of Mayo Foundation for Medical Education and Research. All rights reserved.

green press INITIATIVE Michigan State University Press is a member of the Green Press Initiative and is committed to developing and encouraging ecologically responsible publishing practices. For more information about the Green Press Initiative and the use of recycled paper in book publishing, please visit www.greenpressinitiative.org.

Visit Michigan State University Press at www.msupress.org

This book is dedicated to
P. J. O'Rourke . . . my literary idol and inspiration.
Someday I hope you'll let me buy the first round.

I would like to thank my wife, Julie,
for her hard work and editing expertise.
This is a better book because of her.

Contents

Introduction

Nobody moves to Minnesota for the weather. This observation is as true today as it was on August 21, 1883. The weather in Rochester had been pleasant that morning, but over the course of the afternoon it turned oppressively hot, humid, and hazy. Dr. Will Mayo, twenty-two years old and recently graduated from medical school, had spent the day seeing patients. His eighteen-year-old brother, Charlie, was tagging along trying to make himself useful. The siblings could feel energy building in the clouds overhead.[1]

At 6 p.m. the boys closed the office and prepared to run an errand—a trip to the slaughterhouse north of town, where they hoped to obtain a sheep's head. If the weather cooled down later that evening, young Dr. Will and his doting kid brother were going to practice an innovative new eye operation on the "patient" they planned to bring back with them. Dressed in dark wool suits that made them look more like adolescent undertakers than sophisticated medical men, the brothers set off in their one-horse carriage as the sky began to darken.

Just before the Mayos reached their destination, the massive cumulonimbus clouds roiling overhead abruptly split open. Rain pounded the earth with heavyweight fists. Abandoning their mare, the brothers hustled into a blacksmith shop for protection from the storm. As they ran, the distant sound of an approaching train could be heard over the din of the torrential downpour.

But that was no train.

As Will and Charlie stared westward, a funnel cloud dropped down and began weaving erratically toward them. Moments later the roof of the blacksmith shop tore away, exposing the full fury of the storm. Fleeing the doomed building, the brothers leaped in their buggy and raced back toward town, frantically trying to anticipate the erratic path of the powerful tornado. The Mayos crossed the Broadway Bridge over the Zumbro River mere moments before the wind destroyed it.

The ability of a midwestern tornado to reduce mighty edifices to a state of entropy is no myth. In just over a minute the courthouse, high school, blacksmith shop, and two grain mills were destroyed. Barns were leveled and farm animals killed. Buggies, small fir trees, and long lengths of freshly painted white picket fences cartwheeled across town as if they were auditioning for the circus.

Still, the tornado hadn't finished its tantrum.

Flowing into the Zumbro River was a small tributary called Cascade Creek. As the angry F5 funnel cloud zigzagged its way across the northwest corner of the city,[2] it momentarily dipped into the shallow rivulet, instantly vacuuming the bed dry and turning countless carp and suckers into sushi.[3]

The storm finally ended.[4] Discovering that the town's telegraph still worked, Mayor Samuel Witten sent a message to Minnesota governor Lucius Hubbard telling him "Rochester is in ruins. One-third of the city laid waste. We need immediate help."[5] Twenty-four people had been killed by the tornado, and over forty were seriously injured. As night descended, the living and able-bodied began to pull the dead and injured from their flattened homes. Work was guided by lantern when necessary, but in many cases the numerous fires caused by lightning and toppled lamps provided sufficient light for rescue teams to carry on. Rommel Hall,[6] the city's frontier version of a dance club, had been spared and was quickly converted into an emergency hospital. Thirty-four patients were injured seriously enough to end up there; over 100 others received emergency care outside of this improvised "hospital."

The doctors of Rochester responded immediately. Dr. William Worrall Mayo, the town's best-known practitioner and father of the two young men who had barely survived the tornado hours earlier, made his way quickly to Rommel Hall and began seeing patients. Shortly after the elder Dr. Mayo appeared, another doctor who practiced in the city showed up and ordered emetics for all of the injured. He was rudely overruled by Dr. Mayo, who reasoned that several dozen wounded patients hurling their beans and bacon onto the folks attending to them wasn't going to help matters any. With the potential for infighting to continue, city officials decided they had to appoint a strong emergency director. Recognizing a natural leader in Mayo, they placed him in charge of the disaster response.[7]

The first problem facing W. W. Mayo was that he lacked the help necessary to care for the patients in his makeshift "MASH unit." Knowing that there weren't any trained nurses in town, he asked Mother Alfred Moes and the Sisters of Saint Francis for help. Mother Alfred ran Our Lady of Lourdes Academy, a Catholic primary school in Rochester. Although they were trained as teachers and had no formal medical background, the sisters immediately answered the plea and provided nursing care until all the patients went home.

In the weeks that followed, the city quickly healed. The Mayos were content to return to the status quo of their growing medical practice. Mother Alfred, however, now believed that Rochester should have a hospital of its own. At that time there were perhaps six "real" hospitals in the entire state of Minnesota[8]—two or three in St. Paul, one in Minneapolis, one in Duluth, and a run-down house in Winona where the infirm could stay.[9] These early hospitals were not places you went to get well—the sick usually remained at home, where they were treated by a doctor who instructed family or friends on the proper care of "their" patient. Hospitals were mostly for indigents, a place to die when no one else would care for you. In terms of comfort and hygiene, they resembled college fraternity houses more than modern health care facilities. Mother Alfred envisioned that a hospital could be something more than this, perhaps even a place where high-quality care, sophisticated procedures, and cutting-edge medical innovation would be available. It was a radical concept for the time, but not completely unknown to medicine—in Baltimore, a thousand miles to the east, Johns Hopkins was preparing to open a hospital that would incorporate the same lofty ideals.[10]

Mother Alfred quickly developed a friendship with Dr. W. W. Mayo, and they discussed the possibility of establishing a hospital. The senior Mayo was unenthused.[11] Knowing as much as he did about the nuances of medical practice and the logistical difficulties that Rochester posed to such an undertaking, he doubted it could be accomplished. But Mother Alfred had the kind of persistence that comes with uncompromising conviction. She eventually persuaded Dr. Mayo to pledge his cooperation: he agreed to staff the hospital if Mother Alfred could generate the necessary funds to build it. It took the sisters four years of scrimping to save the $2,200 needed to obtain nine acres west of the city. The land was purchased in 1887, and the hospital opened in 1889 with twenty-seven beds.[12]

The Mayo family name would eventually become a household word, but the contribution made to the clinic's origin by Mother Alfred Moes cannot be ignored. The prospect of a Catholic religious sister building a hospital in Minnesota must have seemed about as likely as a pro wrestler getting elected governor of the state, yet Mother Alfred succeeded despite anti-Catholic bigotry, political rivalries, special interest groups, and an ample supply of Great Plains ignorance. Abetted by the Mayos, she helped a cyclone-devastated prairie speed bump called "Rochester" build a world-class health care facility for itself. The clinic would not only become a medical mecca for patients, but it would also spawn the creativity and cooperation necessary to make this story possible.

Thus were the beginnings of the Mayo Clinic. Dr. Will Mayo, an 1883 graduate of the University of Michigan Medical School, and his brother Charlie, who graduated from medical school in Chicago in 1887,[13] joined their father and began practicing in the new hospital. With the recent discovery of chloroform anesthesia, surgery (the Mayos' chosen field) was poised to make big strides. Antisepsis and cleanliness were emerging concepts, and their implementation was beginning to make operations safer. As infections and other complications began to decline, new surgical techniques were developed for common problems like appendicitis, gallstones, tumors, and trauma. The adage "a chance to cut is a chance to cure" was becoming, for the first time, more than wishful surgical thinking.

In contrast to surgery, the field of medicine was relatively stagnant. Little that was being practiced or prescribed would ever evolve into modern medical therapy as we know it today. Antibiotics, chemotherapy, anticoagulation, Midol, and Botox were decades away. Medicine was the domain of home remedies, patent medicines, bitters, emetics, purges, cathartics, plasters, teas, poultices, bleeding, buckeyes (hung around the neck), and other more-or-less worthless therapies.

The Mayo brothers may have been teetering on the brink of modern surgery, but they were living in the Old Testament when it came to drugs and medicines. Will and Charlie knew nothing of a soon-to-be-discovered class of substances—steroids—that would eventually prove crucial not only to their fledgling clinic but to virtually every aspect of biology and medicine. Of all the ingredients that make up the human body or regulate its function, none are more ubiquitous, historically important, controversial, or economically significant than steroids.

Steroids are complex molecules;[14] they have a skeletal frame comprised of thirty carbon atoms arranged in four interlocking rings, and like the old joke about fans in the stadium on football Saturday in Columbus, Ohio, "they're all related to one another." They include cholesterol, estrogen, progesterone, testosterone, digitalis (a powerful heart medication), and aldosterone (and the other steroids that determine how much salt and water we excrete or retain). And, of course, the focus of this story—cortisone. Steroids may differ from one another by as little as a single atom or chemical bond, but these minor changes in chemical structure have major effects on their actions.

So why are steroids that important? Consider these three reasons:

- First, steroids are building blocks for the body. Just as some houses are constructed from bricks and stones, cells are likewise built of smaller components, many of which turn out to be cholesterol or other steroids. Because of their unique solubility and electrical properties, steroids can form thin membranes that are impermeable to water and other substances. These membranes are the barriers that keep certain things inside your cells, and keep others out. Without steroids to form these barriers and maintain cellular integrity, we'd be reduced to pudding. Other "structural" steroids are important because of their solubility properties—for example, bile acids made by the liver and stored in the gall bladder help us dissolve and absorb food from the intestines. Without them, all we'd be able to eat is pudding.

- Second, steroids help to regulate the function of the body, and in this capacity they are referred to as "hormones." This word, coined just after the turn of the twentieth century by Ernest Starling and his brother-in-law William Bayliss, comes from a Greek term meaning "to excite or set into motion."[15] Hormones "excite" by transmitting messages throughout the body; these messages inhibit, activate, or otherwise modify physiological processes.[16] Proper hormone levels control growth, reproduction, salt and water excretion, and other bodily functions. Hormonal balance produces health, while excesses or shortages of hormones lead to sickness and death.

- Finally, steroid hormones can be used as drugs. It's sometimes obvious how a hormone becomes a drug—for example, if someone has a hormonal deficiency, the problem can be alleviated by giving the missing hormone as a supplement (that is, a drug). That's what happens when someone takes thyroid medicine (a form of thyroid hormone) to correct an underactive thyroid gland. Other applications aren't necessarily so obvious; in fact, when given in unnaturally high doses, or to patients with particular diseases, or in other nonphysiological ways, hormones may exert unexpected, sometimes surprising pharmacological effects.

Drugs generally fall into two categories: those that make you live longer (antibiotics, heart medications, antihypertensives, and so forth), and those that make you live better (analgesics, antihistamines, laxatives, antacids, Viagra, and so forth). Like Aladdin's "Genie of the Lamp," steroids can magically do both. Steroids are the creams that stop skin from peeling. Or make skin more appealing. They are drugs that help a woman get pregnant. Or stop a woman from getting pregnant. Steroids can build hard muscles. Or tenderize meat. They are the vitamins that grow strong bones. Or drugs that grow hair. They help you absorb your food. Or absorb the sun. They are drugs that prolong your life. Or prolong your love life.

Steroids are, without too much exaggeration, the most important pharmaceutical class of all time, and the commercialization of various natural or synthetic steroids has become big business. Thirty of the 300 top-selling drugs are either steroids or drugs that affect the levels of certain steroids; one specific brand of cholesterol-lowering drug (Lipitor, Pfizer Corporation) had recent yearly sales of over $12 billion.[17] And we can't forget the plethora of illegal and non-FDA approved steroids purchased by more than one million Americans at a cost of over $400 million per year.[18]

But the price we pay for something isn't always measured in money, and in the case of steroids the price includes side effects.[19] Common steroid side effects range from mere nuisances (weight gain, poor sleep, facial hair on women, and breasts on men) to bona fide disasters (blood clots, overwhelming infections, or life-threatening potassium disorders). In recent years the social and legal implications of certain steroids have dominated the news, especially regarding performance-enhancing and body-building steroids. The Olympics, baseball, football, cycling, and many other sports have all been tainted by steroid scandals. Unfortunately, this has led to the popular misconception that the term "steroids" implies "anabolic steroids" (testosterone-like substances that promote the growth of muscle and other manly features). It does not, even though more than 80 percent of steroid references on the Internet involve the anabolic variety. In some situations, steroids are taken because of the side effects they produce rather than their primary effects.[20]

It turns out that the various steroid "genies" don't live in magic lamps. They all live together in Pandora's box. The cortisone genie is no exception to this complicated living arrangement. You can let him out and reap the benefits of his magic and power, but not without opening the box and letting out bad things. Expect major consequences.

Forget chemical equations and unpronounceable formulas, because the pharmacology and physiology of cortisone aren't the focus of this book.

The Quest for Cortisone is the story of the discovery of cortisone—or, more precisely, the story of five tightly linked entities at the heart of the cortisone saga: a steroid called "cortisone," Edward (Nick) Kendall, Philip Hench, the Nobel Prize, and the Mayo Clinic. It's a tale with the best literary ingredients—exciting plot, sex and drugs, mystery, adventure, and untimely deaths—told by an impressive and eclectic cast of fascinating (and sometimes seriously flawed) characters. Brilliant scientists. Nervous overachievers. Famous detectives. Insane authors. Troubled artists. Feminists. Saints. Scoundrels.

What is so special about these "five tightly linked entities" around which the cortisone story in *The Quest for Cortisone* revolves, and how do they relate to one another? For starters:

- *Cortisone*—Steroids like estrogen, progesterone, and testosterone may be as familiar to the general public as cortisone (the term *"cortisone"* is commonly used to denote any cortisone-like compound), but they don't have the same life-saving, symptom-relieving, or commercial value of the various cortisone drugs. No other class of steroid is used so routinely and equally by men and women—or children. No other steroids are injected, swallowed, rubbed on the skin, dropped in the eyes, or inhaled as often as the various forms of cortisone.
- *Edward Kendall and Philip Hench*—Kendall, the chemist, isolates and purifies cortisone; Hench, the rheumatologist, finds a use for it. This isn't as easy as it sounds. And it doesn't turn out quite the way anyone imagined.
- *The Nobel Prize*—Kendall and Hench (along with Tadeus Reichstein) win the Nobel Prize for their work on cortisone, but other laureates from various disciplines weave unpredictably in and out of the story.
- *The Mayo Clinic*—The search for cortisone seems doomed to fail on numerous occasions—yet the work is allowed to proceed for decades despite the seemingly dismal prognosis for the project. The extraordinary (if sometimes reluctant) support provided by the famous medical institution thus becomes a critical part of Kendall's and Hench's eventual success.

The origins and history of the Mayo Clinic impact directly on the cortisone story. Doctors Will, Charlie, and W. W., along with Mother Alfred and the city of Rochester, launch the largest, greatest group medical practice in history, and it spawns an environment in which Herculean tasks can be accomplished. The Mayo Clinic of the 1930s and 1940s becomes a place where scientific innovation is encouraged. Diligence in the pursuit of lofty goals is respected and supported. Failure is tolerated. Long-term commitments to science and society are forged—and honored. With a hard, defiant slap to the face of business school graduates everywhere, Mayo becomes a wildly successful enterprise—despite a financially naive core philosophy that suggests doing the right thing might somehow be more important than maximizing profits. This approach makes Mayo the only place where the slow, costly, seemingly futile search for cortisone is encouraged.

The "Mayo approach" also ushers in an era of medical specialization not previously seen. Paraphrasing Adam Smith in his epic 1776 work *The Wealth of Nations*,[21] "Specialization adds value to labor and production." Like Smith, the Mayos recognized that specialization adds value to the practice of medicine. They built their system to take advantage of it. But specialization wasn't the brothers' only contribution to innovation. Basic research, interdisciplinary cooperation, record sharing, open communication— these, like specialization, were also extreme concepts in American medicine before the Mayo Clinic existed, and the consequences of incorporating all of them into a single medical institution turns out to be as powerful as the 1883 tornado that smacked Rochester on the side of its hard midwestern head. By demonstrating how the integration of these ideals under one roof facilitates the discovery of something as important and elusive as cortisone, the value of Mayo's unique approach to science and medicine was established beyond dispute.

There are others who contribute to this tale, and selecting the most worthy of these for inclusion is problematic. If, as some suggest, anything is theoretically related to anything else by a mere six degrees of separation, it's tempting to contrive connections, fabricate relationships, or sensationalize otherwise tenuous links between key elements of the cortisone story and almost any random person or event. To avoid straying too far, *The Quest for Cortisone* focuses on things with one—or in rare cases, two—degrees of separation from cortisone, Kendall and Hench,[22] the Nobel Prize, or the Mayo Clinic. And by doing so a simple conclusion will be reached: success in science—or anything else—isn't guaranteed by genius alone. There are other essential ingredients. Diligence. Tolerance. Faith. Teamwork. And a healthy dose of good luck never hurts.

CHAPTER 1

Addison and His Disease

It is of the highest importance in the art of detection to be able to recognize out of a number of facts which are incidental and which vital. Otherwise your energy and attention must be dissipated instead of being concentrated.

—SHERLOCK HOLMES,
"THE ADVENTURE OF THE REIGATE SQUIRE"

AT TWO O'CLOCK THE AFTERNOON DINNER BELL RANG. DR. THOMAS Addison was ambling through the well-manicured garden outside his home, accompanied as always by two watchful "companions." Unfortunately, the thought of English cuisine again was more than his fragile psyche could handle today; he suddenly broke away from his attendants and hurled himself over a dwarf-wall, diving headfirst onto the stone pavement nine feet below.[1] The impact shattered his forehead, driving jagged fragments of skull deep into the underlying gray matter. The brain damage was massive and irreparable. Dr. Addison never regained consciousness and, according to the *Brighton Herald*, died at 1:00 A.M. on June 29, 1860. His body was returned to his family home in Cumberland, and he was buried in the priory churchyard in Lancaster Abbey on July 5.[2]

Why would a man who had grown up eating haggis kill himself rather than face another home-cooked meal?[3] Two reasons.

First, Addison was depressed.[4] In the mid-1800s clinical depression was poorly understood, largely untreatable, and frequently lethal. Mental illnesses were enigmas (Sigmund Freud, the father of psychoanalysis, was only four years old at the time of Addison's death). The modern discipline of "psychiatry" didn't exist yet. Those around him merely accepted that Addison had for many years suffered from a form of insanity called melancholia, which they believed was caused by overwork of the brain.[5] During the winter of 1859–60 his condition worsened, forcing him to retire from clinical and teaching responsibilities at Guy's Hospital in London. Addison retreated from his London home in Berkley Square and relocated to Brighton, where his family hoped the seashore would restore his faculties. Unfortunately, his depression worsened. After a couple of halfhearted attempts at suicide, his wife and stepson hired two bodyguards, Abraham Quilter and John J. Medcraft, to prevent any further attempts.[6] It wasn't money well spent, as these were the very attendants Addison eluded on his way to fatal humpty-dumptification.

The second reason for choosing death over dinner may have been physical illness. For months Addison had been experiencing abdominal pain, weight loss, and bouts of jaundice—ailments that have led to speculation that he was suffering from gallstones or even cancer of the pancreas.[7] His aversion to food—a symptom commonly seen in conjunction with certain malignancies of the digestive system—might have reflected something more sinister than the body's natural response to English cuisine.

Thomas Addison was born in October 1795 in Long Benton, a rural village of Northumberland near the Scotland border. His parents, Sarah and Joseph Addison, ran a grocery and flour business.[8] As a youth he attended various schools in Newcastle-upon-Tyne. His father wanted him to become a lawyer, but Addison decided to become a physician. In 1812 he entered the University of Edinburgh Medical School.[9] He was proficient at Latin, having mastered it in grammar school, and he prided himself on his ability to take lecture notes in that language.[10] In 1815 he completed a doctoral thesis provocatively entitled "Concerning Syphilis and Mercury" and subsequently received his MD degree.[11]

Medical and surgical therapeutics in nineteenth-century Scotland were embarrassingly primitive, yet in certain areas of biological science—particularly the discipline known as physiology—rapid advancements were being made. Many long-standing mysteries concerning bodily functions were suddenly yielding to the investigations of medical detectives like Addison.[12]

Addison left Edinburgh and moved to Skinner Street in the Snow Hill region of London, becoming a "house surgeon" at Lock Hospital[13] and beginning a short affiliation with St. George's Hospital.[14] After dabbling as a practicing physician, first at the Carey Street Dispensary and later at the Royal Infirmary for Women and Children, he entered Guy's Hospital as a "physician pupil" on December 13, 1817, at which time he paid 22 pounds for the privilege of becoming a "perpetual student." His performance must have been satisfactory; in 1824 he was invited to become an assistant physician, and in 1837 he was promoted to full physician. Together with Richard Bright and Thomas Hodgkin, he would become known as one of the so-called three giants of Guy's Hospital. But Addison's steady rise through the ranks of the prestigious facility reflected his genius, insight, and teaching ability more than his political savvy— indeed, most sources suggest that Addison was, put bluntly, socially challenged.

It's not easy to understand the mind and personality of Addison. His contemporaries described him as withdrawn and unapproachable, traits that hint at the underlying depression that later destroyed him. Some attributed these characteristics to "excessive shyness and sensibility," but that explanation is surely too kind. Withdrawn? Unapproachable? Shy? Hardly. Addison surely intimidated many of those who worked with him; comments abound regarding his "severe, pompous manner," the way in which he chose his words with irritating precision, and a physical appearance that was alleged to strike terror into his students. He may have been hailed by his supporters as a "natural leader," but his detractors portrayed him as a man who could skewer his confronters with a penetrating glance.[15] Given the conflicting descriptions of Addison, one must question whether episodes of subclinical mania may have intervened between bouts of depression—in other words, did he suffer from bipolar disease?

Addison paid a steep price for his eccentric personality. His fame, which was considerable at the height of his career, should have provided him with significant perks, including a large clinical practice of rich patients, scientific recognition, and the other kudos normally awarded to the top medical men of London society. But it didn't. Addison's medical practice was small, and those rewards that trickled down to him—membership in the Royal College of Physicians, invitations to lecture the Royal Society, appointment as court physician, titles, and honoraria—all came later, in many cases decades later, than one would have expected for a physician of his prominence.

Even marital bliss came late to Addison. In September 1847, at the very ripe old age of fifty-two, he married Elizabeth Catherine Hauxwell in the same Lancaster church where he would later be buried.[16] They never had children of their own, although she came with two from a previous marriage.

Addison was the most brilliant diagnostician of the time.[17] As the standard-bearer for the science of medical detection, he meticulously correlated his clinical findings with those noted at autopsy—all the more remarkable considering that autopsy was a rarely performed procedure in the 1800s. In the professionally competitive environment of Guy's Hospital, Addison sought every advantage he could to maintain his mighty reputation as a master of diagnostic annihilation; he readily embraced new diagnostic tools (for example, while others were scoffing at Laennec's recently introduced stethoscope, Addison was championing its use). Like some Hippocratic version of Sherlock Holmes,[18] Addison tackled medical mysteries by rigidly adhering to the code of the sleuth, the first rule of which is to recognize which facts are vital to the diagnosis and which are incidental. Addison believed that the physician's most important duty was to discover the correct diagnosis; therapeutics were an unfortunate annoyance that, as Sherlock Holmes might have put it, caused "energy and attention . . . to . . . be dissipated instead of being concentrated."

Not surprisingly, Addison's patients felt differently: they expected treatment for their ailments. Yet Addison seemed congenitally incapable of devoting as much energy to the cure of his patients as he did to the search for the cause of their disease. As he put it, "Once . . . I have worked out the disease; if it be remediable, nature, with fair play, will remedy it. I do not clearly see my way to the direct agency of special medicaments, but I must prescribe something for the patient, at least, to satisfy his or her friends."[19]

If Addison's temperament terrified his students, he compensated for it with exceptional teaching abilities. He was truly devoted to his pupils, and they to him. As he prepared to leave Guy's Hospital on the way to his tragic ending in Brighton, he wrote a heartfelt letter to the students he was leaving behind, informing them of his ongoing breakdown and thanking them for the devotion they had shown not only to their teacher but to their medical studies as well.[20]

Some of the Addison's "firsts" were major medical milestones. Addison first described pernicious anemia (a form of anemia later shown to be caused by vitamin B-12 and folic acid deficiency); appendicitis; and the various skin conditions caused by diabetes, scleroderma, or high cholesterol. He also was the first to demonstrate that pneumonia was caused by the deposition of fluids and other materials into the alveoli (air cells) of the lungs.[21] He wrote one of the first textbooks of modern medicine (*Elements of the Practice of Medicine*, 1839). These accomplishments, however, are not the ones for which he is best known; it was Addison's work on the suprarenal glands that firmly established his place in history.

While investigating patients with pernicious anemia, Addison fortuitously noted that a few of them had the same unusual constellation of findings. Along with the characteristic anemia and general decline in health shared by all with the pernicious variety, there was also a small subgroup with atypical features that included absent appetite, weak pulse, abdominal pain, emaciation, vomiting, and—most strikingly—an unusual discoloration of the skin. It was the skin finding, a seemingly trivial detail and something only a dermatology expert like Addison was apt to notice, that ultimately convinced him he was dealing with a unique disease entity. The skin typically developed an odd discoloration, which the patients themselves usually noticed. It was described as being dingy or smoky in quality, with tints and shades of deep amber or chestnut brown; in some cases the pigment, which was most prominent on the face, neck, upper arms, axilla, and nether regions, was so dark that the affected patient "might have been mistaken for a mulatto."[22] Addison recognized that a patient with these symptoms gradually weakens and dies. The autopsies he subsequently performed on them revealed something fascinating. And totally unexpected.

The suprarenal glands (later called "adrenal" glands), two small almond-size masses sitting atop the kidneys, had been known to exist since Eustachio described them in the 1500s, but their purpose, if any, was a mystery.[23] Having no discernible function, most medical and anatomical textbooks ignored their existence entirely. By combining careful antemortem clinical evaluation with meticulous dissection at autopsy, Addison ascertained that many patients with the combination of constitutional symptoms and skin discoloration had suprarenal glands that were highly abnormal.[24] Instead of being small and soft, they were "the size of a hen's egg" and as "hard as stones."[25] In the 1800s these changes were most often the result of infection with tuberculosis, an epidemic illness across much of Europe caused by slow-growing bacterial rods with a propensity to attack the suprarenal glands. Based upon his eloquent description of this condition, the syndrome associated with destruction of the adrenal glands was soon referred to as "Addison's disease."[26]

There was just one catch. Addison wasn't the first person to describe "Addison's disease."

Sir Zachary Cope, one of the most eminent surgeons and medical historians of the twentieth century,[27] knew as much about abdominal problems as any doctor before him, with the possible exception of Addison himself. Cope's 1922 monograph, *Early Diagnosis of the Acute Abdomen*, written over 100 years after Addison entered Guy's Hospital as a student, remains among the most influential surgical texts of all time. Right up until his death in 1964 at the age of ninety-three, Cope was solving riddles involving abdominal and gastrointestinal distress. His last "patient" was a woman who had been dead for nearly 150 years when he claimed to have finally diagnosed her problem.

She was Jane Austen, the legendary author. Like many female authors of the time, her novels were published anonymously; despite this, the books written "By a Lady"[28] made Austen one of the most influential novelists of the early 1800s. Her works, which include *Pride and Prejudice, Sense and Sensibility, Emma*, and many others, remain top sellers today. She died on July 18, 1817, at the age of forty-one—a full thirty-eight years before Addison published his description of the condition that carries his name. The cause of her demise has been a topic of interest to medical historians ever since, including Sir Zachary Cope, who, at the time of his own death, believed he had solved the mystery of hers.[29]

There have been many accounts of the afflictions Austen suffered during the last months of her life, but perhaps the best are those that she provided herself through correspondence with her sister, Cassandra, and others.

It is unclear exactly when Austen's illness began, but in July 1816 she began to describe fatigue and back pain. By December of that year her weakness was severe enough to interfere with routine activities. At one point she declined a dinner invitation, citing "the walk is beyond my strength." Her condition waxed and waned. Over the course of the winter her pain and weakness improved, but gastrointestinal symptoms developed. Specifically, she began to experience abdominal pain, nausea, and vomiting. Apparently evoking the ancient Greek humoral explanation for disease, she wrote of her condition that bile was the cause of her suffering.[30]

As spring approached she began to experience new symptoms. Diffuse joint pain developed, becoming especially severe in one of her knees. Her skin also began to change; regarding her complexion, she wrote that it had "been bad enough, black and white and every wrong color." Over the course of the spring her constitutional symptoms worsened, and she began to experience fever and weakness. In the weeks before her death, symptoms such as "faintness and oppression," pain, pallor, and "sufferings" intensified.[31] On her final day she experienced fainting fits that left her too weak to move. Jane Austen died in the arms of Cassandra following an episode of unconsciousness from which she did not awaken.

The detective in Sir Zachary was struck by the similarities between Jane Austen's illness and the disease later described by Dr. Addison. Intermittent but progressive weakness, abdominal pain, nausea, vomiting, weight loss, and, perhaps most important, the discoloration of the skin are all seen with Addison's disease of the suprarenal glands. Although some aspects of Austen's illness seem inconsistent with this diagnosis (for example, her intermittent fevers), these could be explained by the concomitant existence of tuberculosis, which was not only extremely common in her day but, as noted previously, accounted for most cases of adrenal gland destruction. As Cope concluded, "if our surmise be correct, Jane Austen did something more than write excellent novels—she also described the first recorded case of Addison's Disease of the adrenal bodies."[32]

Addison's publication in 1855 of his observations on disease involving the suprarenal glands was the first suggestion that these small, seemingly obscure bits of tissue could have any clinical relevancy. Their importance was confirmed a year later when Charles Brown-Séquard demonstrated that surgical removal of the adrenal glands from animals produced a condition similar to the disease described by Addison—one that inevitably ended in the animal's death.[33] But why? What did these seemingly trivial pieces of spongy meat do that made them so indispensable? What magical substance did they make? What controlled their activity? These questions would not be answered for almost a century.[34]

CHAPTER 2

Introducing Dr. Kendall

They say that genius is an infinite capacity for taking pains.
It's a very bad definition, but it does apply to detective work.

—SHERLOCK HOLMES, *A STUDY IN SCARLET*

EDWARD KENDALL TOOK A SAMPLE OF HIS HIGHLY CONCENTRATED
acid-insoluble thyroid extract and dissolved it in a small amount of ethanol.[1]
He placed the glass vessel holding the aromatic concoction into a steam bath
and stared as the solution began to slowly boil away. It was mesmerizing. For
most chemists, this procedure, commonly called "chemical extraction," was
as boring as a high school production of *Inherit the Wind*. But for young
Dr. Kendall, working through the complex, multistep process for recovering
the mysterious substance in this witch's brew provided just about the most
excitement any human could have without actually sniffing the hot solvents.

It was late in the evening of December 23, 1914; the sleepy twenty-eight-
year-old chemist had been working in the new fourth-floor biochemistry
laboratory of the Mayo Clinic for almost twenty consecutive hours. He
was mentally and physically exhausted. Surrounded by blue-flamed Bunsen
burners, drums of highly flammable liquids, and enough toxic chemicals
to render half the city of Rochester blind, hairless, and impotent, Kendall
knew he needed to remain alert. But he could still not resist the sweet,
seductive advances of Lady Fatigue. He fell asleep on a stool, teetering pre-
cariously over the bubbling mixture as the air around his head was brutally
violated by loud, staccato snores.

When he awoke an hour later the alcohol had evaporated, leaving behind
a white crust surrounded by a yellow, waxy residue. Both substances evoked
his curiosity. Kendall added more ethanol to the beaker, and the yellow

material promptly dissolved and disappeared. Probably just a contaminant, he reasoned. But the strange white crust persisted. Discarding the alcohol and the yellow substance suspended in it, he focused on the insoluble white salty substance lining the beaker. What is this stuff?

Throughout the course of the day on December 24, he repeated the extraction procedure, manufacturing more of the white crust. By the following day—Christmas—he had accumulated enough of the powdery substance to study it. He measured the iodine content, a test of thyroid extract purity, and found it to be 60 percent—significantly higher than the 47 percent iodine compound he'd started with, and far more concentrated than any thyroid extract ever previously prepared. The possibility that he was staring at the world's first sample of pure thyroid hormone crossed his mind. He dissolved some of the crust in more ethanol and poured in a few drops of acetic acid to lower the pH. The snow falling gently outside the laboratory wasn't half as beautiful as the blizzard of delicate white thyroid hormone crystals that suddenly materialized in his flask as the acid was added.

The mysterious substance produced by the thyroid gland had finally been isolated. Edward Kendall, PhD, called the pure crystallized material "thyroxin." When it was later shown to be an amino acid with an amine group, an "e" was added to the end to maintain consistency of scientific nomenclature. It was now "thyroxine."[2]

Kendall had just pulled an allegorical plum out of his figurative Christmas pie, and in doing so achieved one of the great milestones in biochemistry. But amazingly, this wasn't the discovery for which he would be remembered.

Edward "Nick" Kendall was born on March 8, 1886, in South Norwalk, Connecticut. His father was a dentist, and his mother, like virtually every other mother of the time, raised the kids—in her case, Edward and his two older sisters, Florence and Ruth. Both parents were active in the Congregational church and somewhat "puritanical" in their religious beliefs, a quality that was strongly encouraged in the children. Edward was brought up reading the Bible, singing hymns, and praying on his knees every night, and as he noted years later, "The horrors of Dante's *Divine Comedy* and the bliss of heavenly existence were accepted by me as facts not to be questioned." Or at least they were accepted at first. The ever-skeptical Edward wasn't afraid to scientifically challenge the most taboo subjects, including the validity of his religious indoctrination. As a young child, he tested the hypothesis that "God will punish you for swearing," a fate his parents assured him was a painful certainty were the Almighty to ever hear him unleash a foul utterance. One day, alone except for the tall telephone pole against which he leaned for moral support, he bravely muttered aloud, "God damn," half expecting brimstone to strike him dead for this transgression. When it didn't, he realized that the only way to discover his true limits was by constantly testing them.[3]

Kendall enjoyed working with his hands. On weekends he would visit the Norwalk Cast Iron Foundry, a local inferno where a young boy could marvel at the steel-making process, and Miller's Machine Shop, where he would admire the lathes, drills, and other power tools used to create so many intricate items from lumps of cold metal. Inspired to do the same, he would build "things" in a shop his parents allowed him to set up in their attic; he tinkered with telegraphs, induction coils, electromagnets, and other homemade electric "toys." In the eighth grade he told his teacher he wanted to be a "philosopher," although he was unsure exactly what a philosopher was. What he meant was that he wanted to study mechanical and natural phenomena—as he put it, "figure out what makes the wheels go around."[4]

As a teenager, Kendall worshipped his older sister Florence. When she eventually married her beau, Mark Wilbur, Edward couldn't have been happier. Mark was a great athlete, excellent scholar, and a law student at Columbia University. If Columbia was good enough for his idol Mark Wilbur, it was going to be good enough for Edward Kendall. He entered the university in 1904, and decided to pursue studies in chemistry.

His years at Columbia were academically uneventful but socially formative. Edward wasn't a great athlete, but he competed on an intramural four-man shell rowing team. When his fraternity, the mighty men of Sigma Alpha Epsilon, beat their archrivals, Alpha Delta Phi, to win the interfraternity race on the Hudson River, Kendall (the victorious bow oar) received a bronze medal for his team's accomplishment. Recognizing that a first-place finish warranted a gold medal (which hadn't been awarded because of cost), he took his prize to chemistry class and used the lab's electroplating equipment to cover it with gold.[5]

In college Kendall tried to cast aside the heavy yoke of his strict religious upbringing and act like a "normal" college student. The result was, in some ways, the proverbial worst of both worlds. His friends used to say that his religious beliefs didn't stop him from "departing the straight and narrow path" by doing things like playing cards on Sunday, but they "prevented him from enjoying the infraction." Kendall quickly realized that he was a poor social mixer who'd always be more comfortable in the chemistry lab than a social setting.[6]

The chemist obtained a bachelor's degree in 1908, a master's degree in 1909, and a PhD from Columbia University in 1910.[7] Amazingly, he managed to graduate without knowing that iodine was a normal component of the thyroid gland—he missed this question on his final oral exam.[8] After graduation, he moved to Detroit and began working for Parke-Davis Pharmaceuticals.

In the early 1900s Detroit was a city on the rise. The Coney Island hot dog, a deceptively named concoction (it had nothing to do with Coney Island in New York) served with mustard, onion, and all-meat chili, was making its debut in select downtown diners. The Tigers were the nation's best baseball team, led by their sociopathic center fielder Ty Cobb (a .385 hitter whose teammates hated him almost as much as his opponents did). The Model T, introduced in 1908, was rolling off nearby assembly lines in huge numbers, and Henry Ford's "any color you want as long as it's black" philosophy of mass production was beginning to change the way the world thought about travel and manufacturing. Prohibition was still ten years away, and Stroh's beer, the nation's only fire-brewed beer, was plentiful, cheap, and delicious. "Motown" was arguably the soul of America's changing cultural scene in 1910, just as it would become the soul of America's changing musical scene in the 1960s. This increasingly progressive, exciting, fast-growing city was destined to double its population over the next decade. But Kendall was not going to be part of this population surge. He hated his new job—and the city of Detroit. In particular, he found that the free-spirited scientist in him did not like "punching a clock" for Parke-Davis. Whether there was anything more causing his dissatisfaction remains unknown, but after a mere five months on the job he left and returned to New York City.

Back in New York, Kendall began working in the biochemistry laboratory of St. Luke's Hospital, an affiliate of his beloved Columbia University. The bad news: his pay was lousy. The good news: he was allowed to pursue projects of his own choosing, with the caveat that they needed to be "of value to the hospital." On February 1, 1911, Kendall began working on the isolation of a hormone from the thyroid gland.

By 1913 Kendall had taken a simple extract of the thyroid gland (prepared from tons of discarded cattle thyroids) and purified it more than a hundredfold. To confirm the potency of his preparation, he gave it to dogs and measured the change it produced in their urinary nitrogen—the more thyroid hormone, the higher the nitrogen excretion. Unfortunately, he got the distinct opinion that his medical coworkers were unimpressed with his work—their attitude seemed to be "so what?" Even after his extracts were successfully used to treat patients with severe hypothyroidism, he experienced not-so-subtle hostility from the house staff and attending physicians. MDs resented the idea that some PhD was going to show them how it's done, and they were unjustifiably dubious about the potential of biochemical research. A series of other minor insults followed, including an episode in which one of the hospital administrators sent Kendall a box of cereal and told him, in no uncertain terms, to analyze the contents. Kendall threw the letter—and the cereal—into a wastebasket.[9]

In October 1913 Kendall was informed that his salary for 1914 would remain the same as it had been the previous year—a paltry $1,200. He'd been expecting more. The disgruntled chemist had endured enough; he resigned from St. Luke's. "And so ended my efforts to carry out research in biochemistry in an institution that was not interested in research," he later wrote.[10]

Kendall was out of work and needed a job, but he wasn't going to take a thankless, dead-end position again. Where to go next? A friend arranged for Kendall to interview at the nearby Rockefeller Institute with its director, the famous scientist Simon Flexner. At that meeting Kendall enthusiastically detailed his recent work on iodine, thyroid extracts, and their ability to treat goiter and hypothyroidism. Dr. Flexner was clearly unimpressed and made no effort to conceal his disinterest in the applicant. He concluded the interview with a demeaning 1914 version of "don't call us; we'll call you." The young chemist knew that he had been insulted; he "did not like what Dr. Flexner said, and . . . I disliked the way he said it."[11] Kendall didn't get the job. Neither man realized at the time that they would meet again in a few years under ironically different circumstances.

Acting on the advice of another friend, Kendall subsequently contacted Dr. Louis B. Wilson, director of Medical Laboratories at the Mayo Clinic. Mayo's national reputation in medical research wasn't nearly as prestigious as Rockefeller's, but at least Kendall had heard of the place. At the interview he described his work on the thyroid, and not surprisingly Wilson was interested. At that time the Mayo Clinic was becoming known as a premier center in the country for treating thyroid disease, and Kendall's basic science work in this area was the kind of research the clinic could appreciate.

Kendall joined the staff of the Mayo Clinic on February 1, 1914, only a few months before Gavrilo Princip shot and killed Archduke Franz Ferdinand, the heir to the throne of Austria, in Sarajevo. This event would quickly plunge the world into the war to end all wars. If the onset of World War I wasn't distraction enough for a man starting a new job, Kendall opted to complicate his life further by getting married. He and Rebecca Kennedy of Buffalo, New York, were wed on December 30, 1915.[12] In a public display of her commitment to this marriage, the new Mrs. Kendall immediately joined her husband in a part of rural Minnesota that, according to local legend, Iowa had once refused to annex on the grounds that the landscape was too boring.

The thyroid gland was a subject of great interest in the early 1900s. In 1909, while Kendall was still a graduate student, Emil Theodore Kocher, a surgeon in Bern, Switzerland, received the Nobel Prize for his "work on the physiology, pathology, and surgery of the thyroid gland."[13] But interest in the structure and function of the thyroid antedated this by centuries. Paintings by the ancient Egyptians depict women with obvious goiters. Renaissance physicians wrongly argued that the thyroid gland's main function was to moisten the windpipes and lungs. The relationships between thyroid enlargement and the clinical entities we think of as hyperthyroidism, hypothyroidism, cretinism, and myxedema were recognized in the late 1700s. Cancer of the thyroid was described in 1811. Although the function of the thyroid remained an enigma through the 1800s, its role in women was thought to be more important than in men. Some argued it was a barometer of the female mood, engorging with blood and becoming swollen during "irritation and vexation."

Failure to understand the purpose and action of the thyroid did not prevent people from surgically operating upon it, but operations like thyroidectomy (removal of the thyroid) were hellacious undertakings. John Diffenbach, a German surgeon, claimed in 1848 that thyroid surgery was "one of the most thankless and most perilous undertakings."[14] Not wanting to pass up an opportunity to denounce something, in a statement dated 1850 the French Academy of Medicine condemned most thyroid operations.[15]

Yet in some practices of medicine and surgery, progress was being made in thyroid disease. The previously mentioned Nobel Prize winner Emil Theodore Kocher performed thyroidectomies in the late 1800s;[16] in 1880 the average worldwide mortality for thyroid surgery was approximately 21 percent,[17] but in Kocher's series (published in 1884) he described over 2,000 thyroidectomies with a mortality rate of only 5 percent. By the time he was awarded the Nobel Prize in 1909, his death rate for thyroid surgery had fallen to an almost unbelievable 0.2 percent.[18]

The world's other emerging center of excellence for thyroid disease was the upstart Mayo Clinic. The Mayos had recently hired Dr. Henry Plummer (a man whose genius extended to far more than medicine), whose main specialty interest was thyroid disease. He wrote a classic description of toxic nodular goiter, a condition that eventually became known as "Plummer's disease," and he developed techniques for treating patients with iodine prior to goiter surgery; this led to a dramatic reduction in surgical complications.[19] Working with Charles Mayo, the brother who was most interested in thyroid surgery,[20] the operative mortality for thyroidectomy at the Mayo Clinic fell to 3 percent by 1905.[21] Because of the iodine-poor soil in the Midwest, goiters were endemic to this region; once perfected, thyroid surgery became a major activity at the Mayo Clinic. The remarkable growth of the clinic during the early 1900s was a direct reflection of the Mayo brothers' pioneering success in the area of thyroid surgery.

What kind of man had Mayo hired as the clinic's head of biochemistry? Years later, after Kendall's death, Dr. Dwight Ingle, a research associate in biochemistry, wrote a short biographical memoir for the National Academy of Sciences describing the man he often referred to as "Chief."[22] While his tone shows he respected and genuinely liked Kendall, Ingle wasn't reluctant to point out the many quirks and paradoxes that made his mentor's personality so interesting, and his eventual success so remarkable. Like Addison, Kendall had flaws and oddities that should have crippled his ability to succeed as a scientist. Some of these were turned into assets—for example, his compulsive work ethic, unwavering personal discipline, and near-pathological drive to succeed helped him accomplish things that eluded men with lesser intellectual stamina. Other personality quirks remained liabilities. Ingle suggested that "undue faith in his own ideas and resistance to the suggestions of others characterized his whole life as a scientist."[23] With regard to his scientific abilities, "Kendall's knowledge of physiology was shallow: he did not appreciate cause-and-effect relationships, and he did not fully appreciate the extent of biological variability. He was also given to making premature announcements of laboratory results."[24] Perhaps most damning for a serious scientist like Kendall, Ingle perceived him to be "a stubborn man, throughout his life he held that his intuitive beliefs were valued until the evidence against them became overwhelming."[25] And what are we to make of Ingle's two most provocative observations? First, that "Dr. Kendall was a man of scholarly demeanor, but he was not a scholar."[26] And in perhaps one of the great non sequiturs ever, his observation that Kendall "was not then, nor was he to become a great chemist."[27]

Ingle critiqued not only Kendall's scientific ability but also his temperament, noting that his treatment of others was generally kind, but that he would sometimes become sharply impatient and openly sarcastic.[28] Citing one specific example, Ingle describes how "I once caused him to lose his temper . . . and I soon regretted it. He did not raise his voice, but his face flushed, then paled, and his lips and hands trembled."[29]

Conceit? Naïveté? Sloppiness? Stubbornness? Impatience? These aren't traits that would normally advance a scientist's fortunes. But Ingle wisely concedes that "these foibles may have been necessary for his noble aims, his tenacity, and hence his great achievement."[30] That seems logical enough. Maybe those who make great discoveries do so by looking at the same things others have looked at, but seeing them differently. Or by relentlessly pursuing ideas that everyone else knows damn well are stupid and wrong. Or by beating a dead horse until it somehow gets up and walks away.[31]

The isolation of thyroid hormone was a major accomplishment. Had a Nobel Prize not already been awarded to Kocher a few years earlier for his work on the thyroid, Kendall's accomplishment may have warranted a free trip to Stockholm. He still received a modicum of notoriety for his work, and the young chemist took extreme pleasure in presenting his paper, "Isolation in Crystalline Form of the Iodine-Containing Compound of the Thyroid Gland," at the 1916 annual meeting of the Federation of American Societies for Experimental Biology, which was held that year in New York City. The chairman of the meeting, to whom he presented this masterful achievement, was Dr. Simon Flexner—the man who had rudely dismissed him during his previous job inquiry at the Rockefeller Institute. Revenge is a dish best served cold.

Unfortunately, Kendall's efforts to study thyroxine from this point forward failed miserably. He spent the next ten years attempting to synthesize the hormone, but was unsuccessful largely because he had incorrectly concluded that the structure was "triiodo-hexahydro-oxindolepipropionic acid" (his proposed formula assigned three iodine particles to each molecule of hormone; there were actually four). His assistant, Dwight Ingle, could barely find words to describe the jackassish stubbornness with which Kendall stuck to his erroneous guns regarding the postulated molecular configuration. Ingle incredulously noted that "other chemists had advised him over and over that his proposed structural formula for thyroxin(e) was incorrect."[32] But Kendall flat out wouldn't listen.

Kendall ultimately squandered his opportunity to finish the thyroid story himself. In 1926 a Londoner named Charles Robert Harrington, armed with the same information that had frustrated Kendall for a decade, correctly identified the structure of thyroxine, synthesized it, and published his work.[33] At this point, it was believed the thyroid was fully understood, and the Mayo Clinic stopped doing research on this organ.[34]

In his twelve years since coming to Mayo, Nick Kendall had acquired everything a scientist needed to undertake meaningful, high-quality research—a laboratory, adequate funding, a team of talented coworkers and assistants, and the support of a top-notch clinical institution. What Kendall suddenly did not have was a project, direction, or goal.

CHAPTER 3

Life After the Thyroid

What one man can invent, another can discover.

—SHERLOCK HOLMES, "THE DANCING MEN"

WITH HIS WORK ON THE THYROID UNCEREMONIOUSLY USURPED AND finished by rivals, Kendall began contemplating a new adventure in alchemy. Unfortunately, the forty-year-old chemist was trapped in research limbo; a decade of thyroid study had forced him to focus his considerable expertise on an extremely narrow aspect of glandular biochemistry, and like the proverbial "top specialist in the field," he'd learned "more and more about less and less" for so long that it now felt as if he knew "everything about nothing." Kendall's research abilities, already considered suspect by some of his Mayo Clinic colleagues, now appeared less promising than ever; a few even speculated that Kendall's career in research had peaked.

If true, this was terribly unfortunate, because opportunities for new fields of medical investigation were blossoming. The nineteenth century came to an end just as the disciplines of biochemistry and pharmacology were hitting puberty; invention was creating an army of new drugs, while the long-held physiological secrets of people's vital bodily fluids were finally yielding to the process of discovery. Some of these novel substances, like thyroxine, were derived from various glands and organs, while others came from sources so diverse—or bizarre—that one can only marvel at the circumstances leading to their discovery and use.

Now that the thyroxine studies were completed, Kendall wanted to take his research at Mayo in a fresh direction. One of the most promising areas of medical investigation at the time involved newly discovered biological agents produced within the body, many of which had potential pharmaceutical applications. For a thyroid chemist like Kendall, the prospect of conduct-ing research on some of these nascent substances—especially those derived from various glands—was an attractive one. Kendall was surely aware of the fascinating work of this type going on around him, but in this target-rich research environment, what project would he pursue next—especially when his career likely hinged on the decision?

To appreciate the dilemma facing the chemist, and to help understand his eventual choice in its proper historical context, it helps to consider some of the potential glandular research interests that were now competing for Kendall's attention.

Thyroxine wasn't the first important substance to be extracted from a gland.[1] In 1901 John Abel and Jokichi Takamine purified a compound from the medulla (inner) portion of the adrenal gland that exerted a powerful effect on the heart and blood pressure. It became known as epinephrine in the United States and as adrenaline in Britain.[2] (The suprarenal glands thus became known from this time forward as the adrenal glands.) A year later Ernest Starling and William Bayliss discovered a compound produced in the small intestine and released into the bloodstream; it stimulates the pancreas to secrete digestive enzymes and bilelike substances. They called the intes-tinal compound "secretin," and coined the word "hormone" (a messenger chemical that excites the recipient of the message) to describe its function as a signal.[3] In the 1920s Herbert Evans and Joseph Long identified a chemi-cal made in the anterior pituitary gland that controls growth of the body; a deficit of this hormone produces dwarfism, while an excess causes animals and people to grow into giants.[4] Later named growth hormone, it was a compound ripe for further biochemical study.

Although compounds synthesized by the adrenal medulla and pituitary gland offered promising fields of study in the 1920s, an even more inter-esting story was unfolding around another glandular hormone—insulin. The biggest discoveries involving this newly isolated miracle substance had already been made—and a Nobel Prize had just been awarded to the investigators—but considerable chemistry research of precisely the type that Kendall specialized in remained to be done.

Kendall's certain interest in insulin begins with Frederick Banting, a man five years his junior. Banting was born in November 1891 near Alliston, Ontario, and a less likely candidate to win the Nobel Prize in Physiology or Medicine can scarcely be imagined. A poor student with below-average grades, Banting barely passed high school.[5] After graduation he entered Victoria College of the University of Toronto (a divinity school) but reportedly flunked out. Apparently medicine was not as competitive then as it is now; he transferred to the medical school and received his medical degree in 1916. He spent time in the Canadian army (sustaining a severe wound from shrapnel in his right arm), then returned to Toronto for orthopedic surgical training. Although he seemed to be better suited to medicine than the clergy or military, Banting was still not following a path that suggested impending academic greatness—indeed, at that time the qualification for orthopedic surgeon consisted of being "as strong as an ox . . . and twice as smart."[6] Banting opened a private surgical practice in London, Ontario, in 1920 and became an instructor in physiology at the University of Western Ontario Medical School. A year later he joined the faculty at the University of Toronto.

Banting soon became interested in diabetes and insulin. It was already recognized that the pancreas was the source of insulin and that insulin supplementation could theoretically be used to treat diabetes. Previous attempts to administer insulin to patients had been tried; patients were fed large, disgusting quantities of raw pancreas and unpalatable solutions of pancreatic extract. Unfortunately, the insulin in these sweetbread appetizers and pancreas spritzers was too weak to produce much of a biological effect. Why? Because during the extraction process insulin was destroyed by the powerful enzymes produced within the pancreas, rendering pancreatic extracts clinically useless.[7]

Banting's big idea? He wanted to surgically modify an animal's pancreas so that the portion producing destructive enzymes would atrophy, leaving behind (and unharmed) the region of the pancreas that produced insulin. In May 1921 he approached his department chief, J. R. Macleod, with his theory and asked for permission and facilities to pursue the idea. Dr. Macleod teamed him with Charles Best, a first-year medical student, and the two began their work on May 17, 1921.

It was a sweltering summer in Toronto. Willis Haviland Carrier had just patented the "centrifugal refrigeration machine" for cooling buildings, but modern air-conditioning was still years away.[8] Working in the oppressive heat, Banting devised an operation in which the pancreatic ducts of ten experimental dogs were tied off, producing the desired degeneration within the pancreas. Extracts from the remaining pancreatic tissue were prepared. By September 1921 Banting and Best were able to keep Marjorie,[9] one of their diabetic lab dogs, alive by giving her insulin extracted from other dogs.

Macleod understood the importance of this discovery and quickly enlisted the help of James Bertram Collip, a chemist with special expertise in hormone purification. With Collip's help, a new method for extracting insulin was developed; it worked not only with dogs but also with fetal calf pancreases obtained from a local slaughterhouse. As a result, insulin was suddenly available in therapeutically useful quantities.[10]

Once it became clear that insulin could revive dying diabetic dogs, it was inevitable that the hormone would be tried on a human subject. That person turned out to be fourteen-year-old Leonard Thompson. He was near death and in a coma when he arrived at the Toronto General Hospital in January 1922. Thompson, a severe juvenile diabetic, weighed only sixty-five pounds; he was dying from his disease, and even if he regained consciousness he would likely have, at most, a few more miserable weeks of life left. After testing insulin on each other (fortunately there were no immediate side effects), Banting and Best injected the boy with regular doses of their new discovery. Young Leonard started to improve immediately; he became brighter and stronger. Daily injections followed, and he soon put on weight and began to thrive. Thompson lived another thirteen years, dying in 1935—not from diabetes, but from a motorcycle accident and the pneumonia that set in afterward.[11] The human experiment was successful beyond any reasonable expectation, and insulin was instantly recognized as a major medical breakthrough.[12]

The awarding of the Nobel Prize in Physiology or Medicine to Banting and Macleod in 1923 caused sparks to fly. Banting was initially furious; he thought Macleod had done little more than supply him with a laboratory. The Nobel committee disagreed, recognizing that Banting "undoubtedly was the first to have the idea and . . . carry out the investigations," but adding, "it is very likely that the discovery would have never been made if Macleod had not guided him."[13] Banting shared his portion of the award money with Best, while Macleod shared his with Collip, and eventually a tenuous peace was brokered between the four scientific combatants.[14]

Dr. Best eventually became a prominent physician and scientist; his research included pioneering work with heparin, an extract of lung and liver that soon became a crucial drug for preventing and treating blood clots. Meanwhile, Dr. Banting was knighted for his work with insulin in 1934[15] and continued making significant contributions in a wide variety of medical endeavors, including one that would eventually undergo refinement at the Mayo Clinic.[16]

Kendall must have been tempted by all of the unresolved issues involving insulin. Was there a better way to extract insulin? Could it be synthesized from other materials? Could modifications to its structure create more potent, longer-lasting versions of the hormone? There was a tremendous research opportunity here, and Kendall surely considered pursuing this line of investigation.

But two other sets of glands—ovaries and testes—were also yielding exciting, new substances, the most interesting and important of which included:

Estrogens. Dr. Charles Edward Brown-Séquard, the "Father of Endocrinology"[17] (and the previously mentioned author of the 1855 publication on the importance of suprarenal glands and the dire consequences of their removal) continued to dabble with science and physiology into his senior years. In 1889, at the age of seventy-two, he made a spectacular announcement; he claimed that he had "rejuvenated himself" by injecting "sensitive parts of his body"[18] with an extract of guinea pig and dog testicles. Brown-Séquard suggested that extracts made from ovaries might have the same effect on women. A new era of sex hormone physiology and pharmacology was dawning.

The rapid commercialization of sex hormones was predictable. In the 1890s Merck & Company entered the estrogen market with a product called Ovariin. As the name implied, Ovariin was made from dried, pulverized cow ovaries and was sold as a "flavored" powder.[19] Ovariin was the first marketed product offering natural hormones for the treatment of menopausal symptoms. It was an instant success—a surprise considering that the life expectancy of American women in 1900 was approximately forty-five years, which is less than the usual onset of menopause today.

In 1919 Ludwig Haberlandt, an Austrian, showed that drug-induced contraception might be possible; he demonstrated that female deer could be rendered infertile by the injection of extracts from pregnant rabbit ovaries.[20] In 1927 he achieved the same result in mice using the oral administration of ovarian compounds.[21]

During the early 1920s research into the female sex hormones was under way, led by Edward Doisy in the United States,[22] and in Germany by Adolf Butenandt. In 1929 the first estrogen, a steroid called "estrone," was isolated and purified by Doisy; he later won a Nobel Prize for this work. Over the ensuing ten years numerous other estrogen-type steroids were identified. Unfortunately, the early estrogens like estrone and estriol were not soluble in water and therefore had to be mixed with oil and injected through the skin. The administration of estrogens remained problematic until the 1930s, when scientists in Germany and England independently discovered water-soluble estrogens.[23] These were first isolated from urine. By now James

Collip (the chemist from the insulin story) had moved to McGill University and was working with Ayerst Laboratories to develop an estrogen product derived from the urine of pregnant Canadian women.[24] Sold under the name "Emmenin," its main active agent was estriol glucuronide, an agent that was water soluble and therefore could be absorbed if taken by mouth.

Emmenin turned out to be the Viagra of 1930 in terms of skyrocketing sales, but it had faults. It was relatively weak, expensive to make, and had "problems of taste and odor"[25] that seriously threatened its commercial viability. Because of these difficulties, a new source of conjugated (water-soluble) estrogens was needed. This was found in the urine of pregnant horses. Extracts of a pregnant mare's urine (or Premarin)[26] were introduced to the market in the early 1940s, and although it was more expensive (up to five times more) than other estrogens, Premarin quickly grabbed an upscale image and became the consumers' preferred drug[27]—apparently the hormones extracted from horse urine didn't have the same taste and odor issues as those derived from human urine.

At the same time that Ayerst was developing its various extracts, the German pharmaceutical giant Schering, led by its estrogen expert Adolf Butenandt, was making progress with female hormones.[28] The product Schering eventually developed was marketed as Progynon. Faced with the same production problems encountered by the Canadians, the Germans also switched to pregnant mare's urine as an estrogen source and called their new product Progynon II. By 1938 Schering scientists had synthesized ethinyl estradiol, which still remains an extremely popular form of estrogen. Its development substantially lowered the cost of estrogens and ushered in a new era of female hormone therapy.[29]

Progesterones. It turns out there's more to "female hormones" than estrogens. In 1906 two Englishmen, Francis Marshall and William Jolly argued against the "one gland, one hormone" rule by claiming that the ovaries actually produce two separate steroid hormones—one responsible for menstruation and the other for embryo implantation.[30] The first hormone proved to be estrogen. The second became known as "progesterone," and it would eventually play a major role in the cortisone story.[31]

Testosterones. Dr. Brown-Séquard gets credited for postulating the existence of testosterone with his 1889 "juicy liquide testiculair," the gonadal extract that increased his physical strength and mental capacity, improved his constipation, and "lengthened the arc" of his urine.[32] Twenty years later Leo Stanley, the resident physician at San Quentin Prison in California, transplanted testicles from freshly executed prisoners into inmates; many of the recipients claimed that they recovered sexual function as a result.

Similar operations performed by the Russian-French surgeon Serge Voronoff involved the transplantation of monkey testicles into elderly men;[33] the results aped those reported by Stanley.

Not to be outdone by the estrogen hunters of the time, the previously mentioned Adolf Butenandt isolated a few milligrams of testosterone-like material from 15,000 liters of policemen's urine. In 1935 Coroli David and Ernst Laqueur isolated crystalline testosterone from the testicles, and soon both the Schering and Ciba companies were commercially preparing testosterone. Butenandt and Leopold Ruzicka synthesized testosterone from cholesterol and subsequently shared the 1939 Nobel Prize for their work.[34]

Ultimately, Kendall would ignore these hot areas of glandular interest and turn his attention to the seemingly unfathomable adrenal cortex. He'd eventually undertake a project that would consume him for the next two decades. His efforts to study this cryptic gland, about which almost nothing had been learned since the time of Addison and Brown-Séquard, would be met with frustration, bitter disappointment, personal insult, and, above all, repeated, embarrassing failures.

Fortunately, Edward Kendall was the perfect man for the job.

CHAPTER 4

Introducing Dr. Hench

I have a curious constitution. I never remember feeling
tired by work, though idleness exhausts me completely.

—SHERLOCK HOLMES, *THE SIGN OF FOUR*

THERE ARE A MILLION JOKES THAT BEGIN WITH "A MAN WALKS INTO a . . ." We generally assume they are fictitious: the possibility that such an event actually took place sometime or somewhere seems ludicrous, and yet the following story, which has been repeated for decades in various forms as a joke, appears to have actually occurred. Reliable witnesses and independent sources swear this happened—more or less exactly as described.[1] Even if it is not true, it should be.

"A man walks into an elevator . . ." In this case, he's a big, powerful-looking man. The elevator is located in the newly constructed Mayo Clinic main building (later called the Plummer Building); it is a blustery, cold day in the early 1930s. Two men are already on the elevator when he steps in. The first, Joe Fritsch, nicknamed "Joe Clinic," is the genial doorman of the clinic. Normally stationed in the lobby, where he greets and directs incoming patients, Joe is ascending the elevator to run an errand for a pregnant woman. The second man stands quietly next to him. He is Dr. Philip Hench, assistant professor of medicine and head of the newly formed Department of Rheumatology. As he enters the elevator, the big man looks over at Dr. Hench.

"Wer . . . wer . . . where's the library," he stammers.
Hench stares back at him and says nothing.

The hulking visitor speaks again. "Wer . . . wer . . . where's the damn li . . . li . . . library?"

Hench remains silent, a completely blank expression plastered over his face. The elevator stops to let someone else on, and the big man steps off in obvious frustration. "Id . . . id . . . idiot," he mutters angrily at Hench as the doors close.

As the elevator begins rising again, Joe Clinic looks over at Dr. Hench. "You know the library's on the 12th floor," says Joe. "Why didn't you tell him?"

Hench answered without hesitating. "Doooo . . . doooo . . . do you think I wanted ma . . . ma . . . my block knocked off?"

Dr. Philip Hench was "tall and extremely good looking . . . [but he had] a cleft palate [without the associated cleft lip deformity that normally tips a listener off as to its presence]; his speech was somewhat affected, but he worked hard on his elocution and could generally be understood quite well."[2] The elevator story suggests that Hench had a stutter; this is not quite true. The defect in his palate gave his voice a congested, nasal tone and made it difficult for him to form certain sounds. He would often stammer as he tried again and again to get a particular word out correctly. An early surgical attempt to correct his anomaly was clearly not curative; years later his son recalled that "he apparently did have some sort of surgery [to repair the palate defect]—perhaps even some of the earliest such surgery. He must have been quite young, and the story we heard was that he took the train by himself to have it done."[3] According to Mayo Clinic legend, Hench's mother insisted he "ignore this handicap," and that he could "achieve anything" if he did.[4] Those closest to him, including his partners at the clinic, readily conceded that ignoring his speech impediment was something that "he did brilliantly."[5] Put less politely, despite his speech impediment, he was a prolific talker. Put even less politely, at times Hench could be a nonstop chatterbox.[6]

It's unlikely that Hench's safety was ever in real danger during the elevator escapade—Hench was a large man, six feet four and 220 pounds. He looked intimidating. But looks can be deceiving. After lettering in soccer at college, Philip Hench never again demonstrated any aptitude or interest in exercise.[7] He was a man with the athletic prowess of a 1961 Renault. This quirk was well known around the clinic; for example, when Hench retired, his going away party included a "spoof" motion picture in which his son Kahler played the role of his father. The film depicted Dr. Hench exiting one of the Mayo buildings, climbing into his car and driving backward approximately thirty yards to another building, parking the car, getting out, and continuing on his way.[8] Physically underachieving but mentally over-powering, as Dr. Howard Polley put it, seems like an on-target description of Dr. Hench. "His inquisitive mind and perfectionism 'in the matter of learning all of the available facts or opinions about a problem' have been remarked on . . . and this formidable combination of characteristics made him a tenacious researcher."[9]

In reality, Hench's speech impediment was no more detrimental to his career than short stature was to Napoleon's. True, the doctor's speech was imperfect, but he was nonetheless an effective communicator. Hench worked exceptionally hard on his verbal presentations—he wanted the audience to appreciate *what* he had to say rather than *how* he was saying it. He was one of the first Mayo lecturers to use slide projections during his talks; this made it much easier for the audience to capture everything he wanted to get across to them.

Dr. Hugh Butt, one of the best-known physicians and scientists of the Mayo Clinic and a former chair of its board of governors, described Philip Hench as a "lord's man"—a doctor who loved to talk, was great fun to work with, and was extraordinarily good with patients.[10] Indeed, Hench was widely admired for being a relaxed, easygoing clinician, a man who spent long hours working in the library or at home. He was, however, infamous for his late arrivals to work. As Hench himself was fond of saying, "Yes, I come to work late . . . but I make up for it by leaving early."[11]

A lord's man? Probably. A Renaissance man? Certainly. But beyond a few generalities, Philip Hench is not an easy man to plunk into some precise, preconceived pigeonhole. Despite his gregarious nature and love of attention, few people truly knew him. Those who thought they did were probably wrong. "Admired by patients as a warm, sympathetic physician,"[12] he could be as serious as cancer or as frivolous as a ten-year-old boy with a water pistol.

Hench possessed a multifaceted personality, and he excelled at showing different sides of it at different times. Describing him has always been a chore in creative writing. For example, in his citation for an honorary degree as Doctor of Science at Washington and Jefferson College in 1940, he was called a "responsible and honored departmental head in one of the great clinics of the world . . . maintaining a wide and balanced interest in life, in culture, in literature, and in music." A few lines later the same citation goes on to paint him as an "irrepressible clown and insatiable lover of fun, one whose contagious joy goes far in the banishment of fear and disease which even [your] skill cannot cure."[13]

Serious? Frivolous? Either way, Philip Hench was a complex person, and—for better or for worse—complex people inevitably lead complex lives. So how would this speech-impaired gentle giant, this inherently serious but paradoxically "irrepressible clown and insatiable lover of fun," handle the psychosocial complexities that his eventual involvement in the cortisone story would create?

Philip Showalter Hench was born in Pittsburgh, Pennsylvania, on February 28, 1896.[14] His father, Jacob Bixler Hench, was a classical scholar and private tutor who taught the rich and famous youthful denizens of Pittsburgh. His mother was Clara John Showalter, about whom almost nothing of importance has been recorded. Hench attended Shady Side Academy, an elite high school, until his senior year. At that point, he transferred to the University School, a private academy established and run by his father. In the fall of 1912 he enrolled in Lafayette College in Easton, Pennsylvania. Hench always considered himself a "Lafayette man,"[15] so much so that years later this Lafayette pride would set him up to be the butt of a practical joke played on him by his closest friends.

Hench appears to have had the kind of storybook college experience that all undergraduates dream of: he majored in the classics, as had his father, but music and pranks seemed to rule the day. Comments written on his yearbook picture frequently referenced his piano-playing skills, in particular a ragtime version of "Nearer, My God, to Thee."[16]

Hench spent a worried week in April 1914 when he was temporarily suspended from school for a fraternity hazing incident. Under normal circumstances, he might have been expelled; however, in this case there were fifty-four indicted "co-conspirators,"[17] and Lafayette could certainly not afford to get rid of such a large portion of its sophomore class. Hench, along with most of the others, was reinstated.[18]

After graduating from Lafayette, Hench enlisted in the U.S. Army Medical Corp (World War I was on the upswing), but was given a transfer to the reserves and thus was able to resume his education. He enrolled at the University of Pittsburgh School of Medicine and received his medical degree in 1920. (The University of Pittsburgh still presents an annual Hench Award in his honor to a distinguished university alumnus.) Following graduation he spent an internship year at St. Francis Hospital in Pittsburgh. At roughly this time the Mayo brothers began offering a few fellowships to support postgraduate work in medicine. Learning of this unique opportunity to further his training, Hench moved to Rochester and became one of the first young physicians to receive advanced medical education at the Mayo Clinic. In 1923, after two years of fellowship training, he was invited to join the staff of the Mayo Clinic as a first assistant in medicine.[19]

In 1926 Hench founded and became head of the Mayo Clinic's Department of Rheumatic Disease; at that time rheumatic diseases affected roughly three million Americans and were responsible for almost 10 percent of all patient visits to Rochester. Arthritis posed a huge national health problem for which specialty care was in its infancy, and Hench's new Department of Rheumatic Disease offered the first advanced training program in rheumatology available in the United States. Despite his prestigious accomplishments, Hench did not become an instructor in medicine until 1928, and was not promoted to assistant professor until 1932.[20] In 1926, at the time of his precocious appointment as department head, he was the same age as Kendall had been when the chemist isolated thyroxine.

At that point young Dr. Hench was given some tough advice. Someone above him in the Mayo Clinic hierarchy was concerned that his speech impediment might be impairing his ability to interact with patients, and that his problem was potentially severe enough to warrant consideration of an alternative career. Hench was encouraged to go abroad for additional studies in some area of medicine where interaction with patients wouldn't be essential.[21] Just in case. The specialty of pathology was a logical choice for a man with a speech problem (his "patients" are either cut into small specimens mounted on microscope slides or they're dead. Talking to them is purely optional). He spent most of the 1927–28 academic year studying with Ludwig Aschoff, a prominent rheumatology pathologist at Freiberg University in Germany.[22] Hench also spent some time with the famous clinician Frederick von Mueller in Munich. After returning to the United States, Hench obtained a master's of science degree in internal medicine from the University of Minnesota. He completed his studies in 1931 before settling back into his clinical career at Mayo. Fortunately, the concern over his communication skills proved unwarranted. His European sojourn had enhanced his scientific knowledge, but did nothing to thwart his desire—or ability—to remain a clinician. Despite the speech impediment, he continued to practice medicine, satisfy patients, and slowly establish himself at the Mayo Clinic.

While he was becoming accustomed to Rochester, Hench met one of the town's best-known eligible bachelorettes, Mary Geneva Kahler. The

Kahlers were Rochester's most powerful and influential family; they owned hotels, businesses, hospitals, and even a radio station that still broadcasts today (the station's call letters, KROC, were said to mean "Kahler Runs Our City").[23] Hench met Mary Kahler while attending the Rochester Presbyterian Church. When it was learned that the doctor possessed considerable piano skills, arrangements were made for him to demonstrate them to the Kahlers on the church's pipe organ.[24] For the young debutante and heiress, this pipe organ serenade proved to be an irresistible keyboard mating call. Philip Hench and Mary Kahler were married in 1927. They subsequently had two sons, Philip Kahler and John Bixler, and two daughters, Mary Showalter and Susan Kahler. The Hench family moved into a relatively modest Victorian house (modest, that is, by Kahler standards) next to the Presbyterian Church, and raised their children in a low-profile setting.[25]

If not idyllic, family life at the Hench household was at least stable and nonconfrontational. "Our family, like many families at that time, was not prone to airing problems inside the family circle, let alone outside it. From what I have heard, Dad was raised by pretty stern Presbyterians! (Think Garrison Keillor's Lutherans.)" As John Hench put it, "I don't recall many real conflicts with him then."[26]

Boringly mainstream in so many ways, Hench's adventurous side was lived vicariously through his two heroes, one fictitious and one real.

The fictitious hero was Sherlock Holmes. With the help of his wife, Hench assembled one of the most impressive Sherlockian libraries of all time.[27] Among the hundreds of volumes in his possession were many first editions of the various Holmes stories (including four copies of the exceedingly rare *Beeton's Christmas Annual* of 1887, containing the initial publication of *A Study in Scarlet*—Sherlock Holmes's first adventure) as well as British and American first editions, along with translations into dozens of languages. His best friend called him "an admitted nut on the subject."[28] His wife's deep pockets were an obvious aid to assembling such a collection; she gave her husband many of his most cherished volumes as gifts, often bearing touching inscriptions "from Mary to Philip." Hench's eldest son, Kahler, has suggested that his "father's interest in Holmes possibly came from Holmes' perfectionism and deductive mind, traits that might have mirrored [his] father's personality."[29] Whatever the source of this hero worship, it is interesting that Hench was attracted so strongly to someone who was "never wrong." The fear of being wrong would surface again and again in Hench's life.

If Holmes was a fictitious hero, the real hero in Hench's life was Walter Reed, the legendary military physician who explored the origins of yellow fever. It is unclear when Hench's interest in this story began, but it did not seem to crystallize until the 1930s; by the 1940s it was a full-fledged addiction. As he wrote to John Moran (one of the volunteers who survived the initial yellow fever research and eventually became a good friend of the rheumatologist) in 1937: "As a physician particularly interested in medical history, I have been long interested in the story of the yellow fever work in Havana. I have been anxious to get in touch with the survivors of this work. . . . I have been collecting old books on the yellow fever story and I am preparing to do considerable research on the matter."[30]

Although Hench would ultimately become famous for his rheumatology research, he had little formal training as a researcher. His self-directed investigation into the life and work of Walter Reed likely provided some of Hench's most important lessons on how to conduct research.

Over time, Hench would collect thousands of original letters and documents related to Reed's work. But the various works involving Sherlock Holmes and Walter Reed were not the only writings Hench collected: "I am something of a bibliophile and autograph fiend. I have a really nice medical library—original copies of Waltering, Jenner, Beaumont—not many but some. I also have a small but nice collection of letters of Pasteur, Paget, Addison, Lister etc. etc."[31]

Coincidentally, Philip Hench—the man who would intuitively find a way to turn a substance made by Addison's organ (the adrenal gland) into one of the most powerful medicines of all time, and win a Nobel Prize for himself in the process—was collecting original works by Addison long before he had any inkling of his impending involvement in the cortisone story.

CHAPTER 5

Nice Guys, Saints, Eccentrics, and Geniuses

My dear Watson, you as a medical man are continually gaining light as to the tendencies of a child by the study of the parents. Don't you see that the converse is equally valid? I have frequently gained my first real insight into the characters of parents by studying their children.

—SHERLOCK HOLMES,
"THE ADVENTURE OF THE COPPER BEECHES"

THERE ARE PEOPLE CONNECTING THE DOTS BETWEEN CORTISONE, the Mayo Clinic, and the Nobel Prize without whom this story cannot be told. These characters weave in and out of the tale in a manner that initially seems erratic and perhaps even superfluous, but they are ultimately as important to the outcome as any of the main players. Some of these "supporting actors," like Percy Julian, Russell Marker, and Tadeus Reichstein, play obvious, high-profile roles. Others, like Leonard Rowntree, Albert Szent-Györgyi, and the Alvarez family, contributed to these stories in ways that were much more subtle but nonetheless important.

The start of the Roaring Twenties saw the Mayo Clinic moving up a steep growth curve. Armed with Kendall's thyroid extracts, the clinic was becoming a major center for thyroid treatment, including the now-successful thyroid operations that the Mayo brothers were routinely performing. Dr. Charlie took a special interest in thyroid surgery and research, and he was constantly searching for ways to improve the practice. His older brother,

43

Dr. Will, was now at the peak of his career and was widely recognized as a great general surgeon, although he spent increasingly large amounts of time doing administrative work. There was no misunderstanding in Rochester—even before his father's death in 1911, Dr. William Mayo was the brother who ran the business that bore his family's name. Buildings were sprouting up around the campus like spring corn, and there were plans for a massive new state-of-the-art clinic building. This magnificent edifice-to-be was largely the brainchild of Dr. Henry Plummer, physician extraordinaire and right-hand man to the Mayo brothers. The new structure would eventually bear his name—the Plummer Building.

But buildings don't make a medical practice. People make a medical practice. And the clinic was accumulating new people at a record pace. A key recruitment occurred in 1920, when Dr. Leonard Rowntree came to Mayo and became head of the Section of Medicine.[1] Rowntree and Will Mayo were already well acquainted with one another: two years earlier gastric acid had eaten a hole through the wall of Rowntree's stomach, causing a perforated ulcer. Mayo had saved his future partner's life by operating on him when the condition became life-threatening. Lesser men might have died from a toxic combination of food, enzymes, and bacteria pouring out through a hole in their stomach and poisoning the contents of their belly. Rowntree survived. Perhaps he was simply too tough to be put in the ground by a mere gastric perforation—Rowntree was, after all, a product of Johns Hopkins, the prestigious, dogmatic, and intellectually demanding Baltimore medical institution that may have been "born" at the same time as the Mayo Clinic, but now had a hospital and reputation far exceeding that of the upstart facility in Minnesota.[2] Once established in Rochester, Rowntree began recruiting the best physicians from various disciplines to join his new section. Philip Hench, one of his earliest acquisitions, arrived in 1921; he typifies the outstanding type of candidate that Rowntree was attracting to Mayo. An equally impressive recruitment was Dr. Walter Alvarez, who joined the clinic in 1926.[3]

Walter Alvarez was a high-energy clinician and a specialist in the obscure (at that time) field of gastrointestinal motility. He had trained at Harvard, where he'd worked with the legendary gastrointestinal physiologist Walter B. Canon.[4] Instead of going into research after his studies, Alvarez returned to his native California and became a practitioner in San Francisco. He was an extremely popular physician, and his medical practice flourished. But Alvarez was bothered by certain elements of the California medical infrastructure. Not one to hold his tongue when comment was called for, Alvarez publicly condemned certain aspects of the California health care system, and in doing so impugned some of his colleagues.[5]

His attitudes predictably alienated many of those in the medical community—not that Alvarez cared all that much. Expressing a sentiment common to those who achieve success, he noted: "Only the man who never does anything outstanding can go through this life without running into opposition, dislike, and bitter criticism."[6]

Alvarez began to search the help-wanted ads in medical journals for a new opportunity, something that would let him conduct research and practice in the manner he wanted. After a chance meeting with William Braasch, a member of Mayo's board of governors, his interest in the clinic was piqued.[7] One can't help but wonder what cards Rowntree was holding when he persuaded Alvarez to leave sunny California and move to the permafrost of Minnesota.

Alvarez's demeanor didn't change much after he arrived in Rochester. He was considered "nice as hell," although some of his theories about medicine differed strikingly from the conventional wisdom at Mayo. He had the idea that "crazy symptoms" could be due to organic disease as opposed to "chronic nervous exhaustion," a view not shared by most of his colleagues.[8] He also believed that stress, troubles at home, and food allergies could cause physical problems.[9]

Not surprisingly, a lot of people secretly made fun of Alvarez; some thought he was "a fussy old man who wasn't such a good doctor."[10] But as the former head of the clinic's board of governors ultimately concluded, "In reality, he was just ahead of his time."[11]

Walter Alvarez's family had branches like a mighty oak tree. His mother came from a prominent family in St. Paul, Minnesota, giving the Alvarez clan strong geographical roots around the Mayo Clinic. His father, Dr. Luis Alvarez, was a Spanish-born physician from the Asturias region.[12] Dr. Luis Alvarez found it impossible to follow a simple path through life. After graduating from Cooper Medical College (now Stanford University School of Medicine), Luis decided to take a short break and visit Hawaii. That "short break" lasted almost two decades.[13] Just after the turn of the twentieth century, when Hawaii fell under the protection of the United States, Luis "felt that his work there was done"[14] and decided to return to California.[15] In 1908 Alvarez moved his family into a Victorian mansion in Los Angeles near the campus of the University of Southern California.[16] Walter was the eldest of five children; Mabel was the youngest (and, in many ways, the most remarkable of the Alvarez offspring).[17] Financially secure, Dr. Luis Alvarez spent the rest of his life (until he died in 1937) taking care of predominantly poor Mexicans living in Los Angeles.

Walter Alvarez joined the Mayo Clinic, and, like many Mayo doctors, he was expected to undertake research as well as to see patients. He did this for seven years, then lost his laboratory space and protected research time as a result of financial cutbacks mandated by the Great Depression. Other doctors, including his boss, Leonard Rowntree, became disgruntled when their research space was taken away. Not Walter Alvarez. The gentle-looking, bespectacled physician cheerfully resumed seeing patients full time, as economic conditions now mandated, and became an expert on "digestive neurosis."[18] He was widely recognized for his practicality and the skill with which he handled patients who were incurable or—even worse—difficult and demanding.

Walter Alvarez quickly became a friend and confidant of Philip Hench, one of his closest clinical colleagues, and developed personal ties to his boss, William Mayo. Mayo was impressed with the new doctor's insight, honesty, enthusiasm, and dedication—but, one suspects, not Alvarez's diplomacy. Alvarez never hesitated to vent his frustrations to Dr. Will over the clinic's inefficiencies, especially the endless committee structure that plagued the clinic starting in the 1920s. Alvarez thought bureaucracy was a major impediment to progress: "A man will surely have trouble as I did if, when something urgently needs to be done, he does it without waiting until two or three committees have discussed it at length and decided to block it."[19] Alvarez's view on institutional bureaucracy seemed to be channeling Hench's hero, Sherlock Holmes, who once observed that "there is so much red tape in these matters."[20]

Walter Alvarez arrived in Rochester with his family, including his fourteen-year-old son Luis. Named after his Hawaii-loving physician grandfather, Luis Alvarez was as disinterested in medicine as his father and still-practicing grandfather were enthralled by the field. Young Luis, much like young Edward Kendall, was more interested in electronics and machinery. By the age of ten, he was proficient with the tools in his father's shop and had a keen understanding of electrical circuits. He found the Rochester High School science classes "adequately taught [but] not very interesting";[21] in retrospect, they were too easy for him. At a time when other kids wore out their eraser before their pencil, Luis used a fountain pen on his math exams. He rarely made mistakes. But if Luis seemed uninspired in his studies, Rochester nonetheless had a positive social effect on him. As he noted years later, "If I had remained in San Francisco, I think I would have been a different person. In Rochester I came out of my shell."[22] In response to Luis's growing interest in physics and engineering (his interests in all other aspects of high school were waning), Walter Alvarez hired a Mayo Clinic machinist to give his son private lessons on the weekends and during the summer. The teenager soon found himself working part-time in the machine shop at Mayo.

Luis was not unlike many adolescents then—or now. He experimented with a "controlled disrespect for authority," and would later suggest that this behavior was crucial for his ultimate development as a scientist.[23] Crucial or not, it must have given his parents—who were surely anxious to be accepted into their new conservative medical community—fits and spasms. While pursuing his "controlled disrespect," Luis engaged in shenanigans worthy of "that irrepressible clown," Dr. Hench. For example, it was not uncommon for Luis to sneak onto building construction sites in the middle of the night and "investigate" the ongoing activity. In May 1927 (while Charles Lindbergh, another Minnesotan, was making his famous transatlantic solo flight) Luis crept onto the fenced-in Mayo Plummer Building construction site; at that time the embryonic edifice was little more than a towering skeleton of steel beams. He climbed the metal framework to its summit almost twenty stories up. On another occasion the reckless young trespasser scaled the 200-foot-tall Franklin Heating Station in downtown Rochester—a gymnastic feat perhaps worthy of Dr. Kendall's Manhattan Bridge crossing escapade.[24]

Luis Alvarez left Rochester to become a freshman at the University of Chicago in 1928 and rarely returned after that. His destiny—which turned out to be quite remarkable and would give hope to frazzled parents of incorrigible children everywhere—lay in another direction.

Kendall's kids were learning their limits, too. Some of them were, like Luis Alvarez, discovering the joys of climbing the Plummer Building. Luis had done it in 1927 when the building was still under construction. But in 1935 Nick's son, thirteen-year-old Norman Kendall, decided that he and a friend needed to "touch the light" on the roof of the building.[25] Sneaking onto the roof by climbing out the fourteenth-floor window of the carillon, the barely-a-teen scaled a tall, narrow metal ladder to the beacon and grasped it, hanging like King Kong atop the Empire State Building, before returning to the safety of the bell tower. Norman's older brother, Roy, a gifted teenage craftsman who once built his own kayak, stretched daredevil to a new level one day by actually maneuvering out onto one of the stone griffins protruding from the top of the Plummer Building; he rode on its back like a cowboy Quasimodo watching people scurry, like little ants, on the street below him.[26]

Hugh Kendall, the third son, had childhood experiences that were different from those of his brothers. And more sobering. Hugh was in sixth grade (the same age at which brother Norman tempted fate at the top of the Plummer Building) when a cave-in killed a classmate who was playing in a construction ditch next to her house. His peer's untimely, tragic death upset him, and "the horror stayed with him a long time." He became obsessed with death, spending sleepless nights "haunted by the sight of the dead girl" who had been buried alive. He eventually overcame his fears by taking advantage of a friend's family situation; the boy's father owned a funeral home in Rochester. Hugh made visits until "I got used to seeing dead people in caskets. . . . Finally I stopped having nightmares."[27]

Kendall may not have been a "great chemist" (at least according to Ingle), but a man named Russell Marker certainly was. Part genius, part eccentric, and in sharp contrast to the stereotypical image we have of most chemists—tireless titrators toying with their trivial test-tube tribulations—Russell Marker was just a bit of a madman.[28]

Russell Marker was born on the family farm outside of Hagerstown, Maryland, on March 12, 1902. He attended the University of Maryland, receiving his bachelor's degree in 1923 and his master's degree in 1924. He was set to receive a doctoral degree in chemistry in 1925 (not bad for a twenty-three-year-old), but in order to graduate he was required to pass a course in physical chemistry. Having just completed his master's in physical chemistry, Marker thought it was pointless to take another course in an area where he was already an expert. The university disagreed and insisted he complete the requirements. Marker's thesis adviser, Morris Kharasch, urged him to take the class and get his doctorate. If he didn't, Kharasch threatened, he'd become a "urine analyst" or "urine boiler." (To a serious chemist, "urine analyst" is the ultimate dead-end career.)[29] Marker was undaunted; he quit without getting his doctoral degree and set off looking for a research job. The cream eventually rises to the top. Despite his painful departure from Maryland, Marker was able to land a job doing hydrocarbon research for the Ethyl Corporation. His work led to discoveries that helped reduce engine knock, and in doing so he developed the octane rating system for gasoline that is still used today.[30]

Always restless, Marker's interests changed after a few years.[31] In 1928 he parlayed his success at Ethyl into a position at the Rockefeller Institute (ironically, it seems he got the job for which Kendall had been turned down fourteen years earlier). He published more than thirty papers during the next six years, most of which dealt with the light-refracting properties of various molecules. It was during this time that Marker became interested in steroids. When he attempted to switch his career to this area, his superiors at the Rockefeller Institute refused. Marker, a man with a low threshold for frustration and a high level of wanderlust, quit Rockefeller and moved to Pennsylvania State University, where he accepted a position funded by Parke-Davis (the same pharmaceutical company for which Kendall had worked more than a decade earlier). The year was 1938, and Marker had found a home in the shadow of Mount Nittany. A temporary home.

The next major actor in the cortisone story is Tadeus Reichstein. He was born in Wloclawek, Poland, on July 20, 1897, and grew up in Kiev, where his father worked as an engineer.[32] Reichstein's early education was obtained at a boarding school in Jena, Germany, and later in Zurich, where he entered the state technical college in 1916 and graduated in 1920.[33] He completed his PhD there in 1922, and continued to work with his professor (H. Staudinger) on the biochemical aspects of aromatic substances in coffee and chicory.[34] His work was financed by various industrial concerns and continued for almost a decade.[35]

In 1938 Reichstein left Zurich, moved to the University of Basel, and became the professor and director of the pharmaceutical institute located there. Just as Russell Marker's academic position at Pennsylvania State University was financed by Parke-Davis, Reichstein's new position involved close ties with Ciba Pharmaceuticals. Reichstein was now well positioned academically and well funded by chemistry standards,[36] and he was poised to make a Nobel contribution that would equal that of Kendall or Hench.

The cortisone story requires the introduction of two more key scientists at this time. The first is Percy Julian. In some ways his story provides more insight into the character of his family than it does into the character of the man himself.

When Percy Julian was born in 1899, America was already a far more progressive place for blacks than it had been for his ancestors. As he stood upon an Alabama railway platform on a warm fall day in 1916 awaiting the train to take him to a fresh beginning at DePauw University in Indiana, Percy could not help but marvel at the opportunity he faced. College. His father, James Sumner Julian, had managed to become a railway clerk in an era when African Americans were not even allowed to ride on the same trains that they served. For many, railway clerk seems a like mundane job; for a black man in the late 1800s, it was a remarkable position. But the Julians were, after all, a remarkable family.[37] Many of Percy's kinfolk had served in the forefront of the uphill battle for social equality, and on this momentous day some of the bravest soldiers from that war were still around to see him off. There, standing on the station platform along with the rest of his family, was Percy's ninety-nine-year-old grandmother—herself a former slave—and Percy's grandfather.[38] Two fingers were missing from the old man's right hand; they had been cut off half a century earlier as punishment for violating the unwritten code that forbade slaves from learning to read and write. It seemed that Julian's destiny had been prepaid by the sweat, indignation, and blood of his forbearers.

There was no doubt that Percy Julian would succeed at DePauw. His parents had already seen to it. The software for success was preloaded into his DNA.

The final scientist to join the cortisone story at this juncture is Albert von Szent-Györgyi. Born in Budapest on September 16, 1893,[39] his father was a wealthy landowner, and his mother came from a famous medical family; her father and brother were both professors in the medical school at the University of Budapest. Szent-Györgyi had the Hungarian good looks and presence of Liberace—but without the flamboyance. He began medical school in Budapest, but his studies were interrupted by World War I. Szent-Györgyi entered the army, fought in Italy and Russia, and won the silver medal for valor before being wounded in action (or, as some claim, shooting himself in the arm);[40] he was discharged from the army in 1917.[41] Szent-Györgyi came to Cambridge, England, as a Rockefeller fellow, and in the fall of 1929—precisely the same time that Hench was off in Europe studying rheumatology and pathology—Szent-Györgyi came to Rochester to work in the laboratory of Dr. Edward Kendall.

Szent-Györgyi had written to Kendall earlier asking if he could temporarily join the Mayo staff (in biochemistry) as a visiting scientist while he was in the United States.[42] He had a specific plan. He had recently isolated several grams of an interesting material called hexuronic acid from the adrenal glands of cows. Although he had also found this same chemical in many fruits and vegetables, its highest concentration seemed to be in the adrenal cortex. He hoped to come to Minnesota, obtain large quantities of adrenal glands from the meatpacking houses in the area (particularly those in St. Paul), and from this make a large quantity of hexuronic acid with which to conduct additional studies. Kendall agreed to host the Hungarian.

Szent-Györgyi's eight-month visit to Rochester would set the quest for cortisone on its way and forever change the lives of everyone involved.

After spending ten years on a failed effort to synthesize thyroxine, Nick Kendall was still in search of a new chemistry project to undertake. Although the structure of thyroxine had been elucidated by others, its mechanism of action in the body remained a mystery. Exactly how did thyroxine affect metabolism? Why did a deficiency of thyroxine turn normal humans into slow-moving, thick-skinned, hairless "toads," while an excess of thyroxine created sweaty, trembling, hyperactive Tasmanian devils? Kendall's knowledge of physiology was limited, but to him, "It seemed possible that a study of the oxidation of *cysteine* and *glutathione* would help explain the influence of thyroxine on oxidation in the body."[43]

It was a shaky physiological rationale upon which to base a major laboratory commitment, but nonetheless in 1926 Kendall decided to embark on a chemical investigation of glutathione, a proteinlike substance that at the time was known to exist but had an undetermined chemical structure. He wanted to discover its composition and formula. In 1929, after three years of clever experiments and some solid organic chemistry, Kendall and his coworkers correctly identified glutathione as a compound made up of three separate amino acids—cysteine, glutamine, and glycine.[44] In the great scheme of medicine, it was a relatively minor milestone.[45] But in chemistry circles Kendall's accomplishment was recognized as good science worthy of accolades; more important, it was a badly needed shot in the arm for the chemist after the thyroxine formula debacle.

CHAPTER 6

1929 and the Decision to Hunt for Cortisone

You know my method. It is founded upon the observation of trifles.

—SHERLOCK HOLMES, "THE BASCOMBE VALLEY MYSTERY"

THE GREAT DEPRESSION KICKED OFF WITH THE STOCK MARKET crash on Black Tuesday, October 29, 1929, and as a result most of the civilized world was soon feeling depressed.

But not Ernest Hemingway. Life in 1929 was going surprisingly well for the testosterone-stoked writer. "Demon Depression," which typically followed him as surely as the darkness of night follows day, could only bide its time for now. The future Nobel Prize winner had just published an enormously successful novel, *A Farewell to Arms*. At the University of Chicago (just a few miles from Hemingway's place of birth and boyhood home), an overworked college underclassman from Rochester, Minnesota, named Luis Alvarez was finding the time to read it.[1]

Nor was depression an issue in 1929 for another man with Nobel connections. In fact, 1929 was turning out to be the best year of Frank B. Kellogg's life. Something was happening to Kellogg that would soon plunge the little collection of farms known as Rochester, Minnesota—and the growing medical clinic that sat in its midst—into the international limelight, just as cortisone was destined to do decades later.

Frank Kellogg was born in Pottsdam, New York, on December 22, 1856.[2] After the Civil War ended, many eastern families migrated westward; the

Kelloggs were one of them. They moved to the small southeastern Minnesota town of Elgin in 1865 and took up farming. Details regarding Frank Kellogg's education are sketchy; he attended a country school, but it is unclear whether he ever graduated. Around 1870, roughly thirteen years before the town would be flattened by the great tornado, he began working at a small law office in Rochester, where he taught himself law along with history and the classic languages. Kellogg passed the state bar examination in 1877 and began practicing law. He was the city attorney for Rochester for two years, and later became the attorney for the rest of Olmsted County.[3]

Kellogg was elected to the U.S. Senate in 1916, just in time to participate in America's entry into World War I. His run for a second term in 1922 ended in defeat. But having tasted the sweet fruit of national politics, Kellogg was determined not to allow a mere election loss to remove him from public service. He left the Senate and immediately became a professional diplomat under President Warren Harding. After Harding's death in 1923, Kellogg was named ambassador to Britain by Calvin Coolidge, and later served as Coolidge's secretary of state.

From 1927 to 1929 Kellogg worked on "the adoption of a multilateral treaty renouncing war as an instrument of national policy,"[4] which was eventually signed by sixty-four countries. Described by one senator as an insignificant "international kiss,"[5] the treaty was predictably broken by armed conflict within a few months of its completion. Frank Kellogg's efforts were, nonetheless, recognized in Norway and deemed worthy of the 1929 Nobel Peace Prize.[6]

Incredibly, the tiny city of Rochester, with a population of just a few thousand, had produced a man who, despite his lack of formal education, became a U.S. senator, secretary of state, and the first of five Rochester-affiliated Nobel Prize winners. The population marveled over the unimaginable way in which the goddesses of good fortune and fame had just kissed the tiny Minnesota hamlet squarely on the lips. What no one realized in 1929 was that they had more sweet kisses to bestow.

Albert von Szent-Györgyi arrived at Mayo in the fall of 1929 and hit the ground running. Quickly obtaining a wooden press, large meat grinder, and numerous forty-gallon crocks,[7] he began the demanding process of extracting hexuronic acid from the seemingly endless supply of bovine adrenal glands that were now available to him from the stockyards of St. Paul. It was messy, tedious, mindless toil; nevertheless, the Hungarian tackled it with the passion of a man who realizes he has been given a rare, one-time opportunity to do something extraordinary. But things changed dramatically when the workday ended. Outside the laboratory Szent-Györgyi was an insatiable social butterfly; he found the means to spend as much time partying and making friends as he did laboring over his crocks of crushed cow glands.[8] He also enjoyed chess, a passion shared by Edward Kendall, and he introduced to the lab a form of four-handed chess he'd learned in England; games took place at least once a week, and sometimes more often. Szent-Györgyi's visit to Kendall's laboratory was, if nothing else, becoming an enjoyable Minnesota vacation for the Hungarian.

While Szent-Györgyi was "borrowing" Kendall's lab to enhance his collection of hexuronic acid, some very serious work on the suprarenal cortex was taking place in New Jersey. Specifically, two Princeton scientists, Wilbur Willis Swingle and Joseph John Pfiffner, were preparing extracts of the adrenal cortex—extracts that seemed to contain a substance that was necessary for life. Frank A. Hartman of the State University of New York at Buffalo was also making extracts of the adrenal cortex; he began calling the yet-to-be identified essential ingredient "cortin."[9] The name stuck, and for the next two decades scientists probing the activities of the adrenal cortex would describe their work as "the quest for cortin." Swingle and Pfiffner demonstrated the physiological potency of their particular solution (which was extracted from beef adrenal glands) by testing it on cats that had undergone bilateral adrenal gland removal.[10] As long as the solution was available, the glandless cats lived. When it ran out, they died. This specific quality—the ability to keep an adrenalectomized animal alive—was the essence of cortin, and for scientists interested in Addison's disease, the creation of these extracts was a stunning accomplishment. Cortin, it seemed, truly existed and could potentially be isolated.

Stunning? Why? Because the mortality rate for Addison's disease (failure of the adrenal cortex) in humans had traditionally been 100 percent. No one afflicted with it ever survived.

But this dismal outcome was being challenged by doctors at the Mayo Clinic who were now treating Addison's disease with the "Muirhead regimen," a therapeutic approach based on the ingestion and injection of adrenal glands and their extracts.[11] The therapy tended to help, but "the untoward results which may appear from the hypodermic administration of epinephrine are a feeling of general weakness and tremor, with palpitation or pounding of the heart; from its rectal injection, tenesmus; and from the employment of the whole gland or cortex by the mouth, gastric or abdominal distress, nausea, vomiting, and intestinal cramps."[12] However, when taken in conjunction with rest, exercise, diet, outdoor life, and sunshine, patients with Addison's disease survived longer. It was a less-than-perfect treatment, but a treatment nonetheless.

Mayo's involvement with the Princeton cortin researchers began in 1929 when Dr. Pfiffner came to Rochester to attend a scientific conference and to visit with Leonard Rowntree. At that time Dr. Rowntree had a special interest in Addison's disease; he was responsible for seeing most of the Mayo Clinic patients with this condition, and he oversaw the Muirhead regimen treatments that many of these patients received. Rowntree was fascinated by the extract Swingle and Pfiffner had produced, and he obtained some of it from them for his own use. He wanted to test it on humans with Addison's disease, something the Princeton group could not do because they lacked a medical school or research hospital.

Writing in the August 6, 1930, edition of the *Staff Meetings of the Mayo Clinic*, Rowntree described the first human to be treated with the Swingle-Pfiffner extract.[13] The patient was a thirty-nine-year-old farmer from Iowa who had been sick with a mysterious illness for several years, and during the past year he'd developed brownish pigmentation over most of his body. The man suffered episodic nausea and vomiting, weight loss, and bouts of extreme warmth. He arrived at the Mayo Clinic in "a critical condition"— his blood pressure was 78/40, and he was severely dehydrated. A diagnosis of Addison's disease was made. He was immediately placed on the Muirhead regimen, and after thirty-nine days he'd improved so much that he was able to return home. As had been the case with most other patients, his symptoms eventually returned and progressed. When he came back to the clinic for a second time, he was in a "state of collapse."[14]

This time the Swingle-Pfiffner extract was ready (it had arrived from Princeton by air on the sixth day after the patient's admission to the hospital). Within thirty-six hours of the initial administration of this substance, there was a marked effect on his strength and appetite.[15] A price was paid for his improvement: severe irritation occurred at the injection site. Eventually the treatments were discontinued because of the intense local pain that the injections were causing (the pain was thought to be due to epinephrine that contaminated the solution).

Supplies of the drug were limited, and when it ran out the symptoms of Addison's disease returned. Despite all the uncertainties surrounding the use of adrenal extract solutions, forty-eight Mayo Clinic patients were treated with this compound between 1930 and 1934.

But clearly a better form of treatment was needed.

Szent-Györgyi finished his brief stay at the Mayo Clinic in 1930 and returned to Europe, where his career took off in a way Kendall described as "meteoric."[16] While in Rochester he had successfully isolated several grams of hexuronic acid as planned. But Szent-Györgyi had no idea what—if anything—his new compound did.

One particular follow-up experiment seemed obvious. Because hexuronic acid was found in the adrenal gland, there was a chance that it could be the mysterious "cortin" for which Kendall and others were searching. To test this possibility, hexuronic acid was given to adrenalectomized animals.[17] They all died, thus proving that hexuronic acid wasn't a substitute for the essential substance manufactured by the adrenal gland; it therefore could not be cortin. So what was it? Szent-Györgyi wondered if hexuronic acid might be related to another newly discovered compound of great interest— vitamin C.

A scientist named Casimir Funk had coined the term "vitamin" (a substance essential for life), or "vitamine" (because its chemical structure contained an amine group), while searching for thiamine (vitamin B1) in the early 1910s; since then, the hunt for new vitamins had become a popular focus of research. Investigators knew that certain fruits and vegetables must contain one of these elusive "vitamins," since some unknown substance present within them could prevent scurvy. Once identified, it would become the third vitamin discovered (vitamins A and B had already been characterized), and it was therefore referred to as vitamin "C." When preliminary investigations conducted by a friend of Szent-Györgyi suggested that hexuronic acid was not the elusive vitamin C, this line of research was abruptly discontinued.[18]

Szent-Györgyi realized he wasn't going to be able to study the biological properties of hexuronic acid; he was not trained in physiology. His strength—and interests—lay in other directions. As he left Mayo, he gave Kendall samples of hexuronic acid and invited him to "study it in his laboratory" if he wished.[19] Kendall decided to make some additional hexuronic acid himself (he extracted approximately 10 grams from his supply of adrenal glands), but like Szent-Györgyi, he also lost interest in the compound.

This was not, however, the last the world would hear of hexuronic acid.

Having lost the ability to conduct research in his own laboratory, Rowntree became discontent at Mayo. One can only imagine how jealously he must have eyed Kendall's laboratory, which, thanks in part to the recent visit by Szent-Györgyi, was continuing to grow and flourish despite the Great Depression.

Kendall's work on glutathione had restored some of the chemist's luster. It was time for him to pursue a new project. Although Szent-Györgyi was gone, the lab was now fully equipped to handle studies involving the adrenal gland. Kendall addressed the situation with Rowntree. After some friendly discussion about the direction the lab should take, Rowntree suggested that Kendall might consider changing his focus to the suprarenal glands. Specifically, why not pursue the identification and purification of "cortin"? Kendall concurred.[20]

Rowntree became increasingly frustrated over the loss of his lab, and, as usual, he did not keep his feelings to himself. Relationships with his colleagues in general—and with his boss, Dr. William Mayo, in particular—deteriorated rapidly.[21] Rowntree resigned from the clinic in 1931 with a big chip on his shoulder. Many of his clinical duties fell to the partners that remained on staff, including Walter Alvarez, who happily gave up laboratory experimentation (a costly activity that was draining the clinic's coffers) in order to facilitate the clinic's goal of seeing more new, money-generating patients.

Less than a year before the death of Doyle, his literary idol, Hench made an observation worthy of Sherlock Holmes.[22] The rheumatologist noted, for the first time, that one of his patients with rheumatoid arthritis experienced a remission in his arthritis during an attack of jaundice.[23] Until now, rheumatoid arthritis had been thought to be an irreversible, inevitably progressive condition, but Hench's subtle 1929 observation (a true "observation of trifles," as Holmes once bragged) raised the possibility that the destruction of human joints might not be inevitable or irreversible in rheumatoid disease. Employing Holmes-like logic, Hench took his observation a step further and deduced that an "anti-rheumatic substance X" must be produced by the body during bouts of jaundice.[24] He reasoned that this substance was responsible for the arthritic remissions; if it could be isolated, substance X might prove useful for treating patients with rheumatic diseases. Over the next nineteen years he would confirm his initial observation countless times by identifying other patients with rheumatoid arthritis who improved after developing jaundice; each encounter encouraged him that he should search for this undiscovered pathophysiological substance hiding within his patients.

Amazingly, Hench would find his mysterious substance X.

CHAPTER 7

Another Kendall False Start, Another Great Announcement

Perhaps when a man has special knowledge and special powers like my own, it rather encourages him to seek a complex explanation when a simpler one is at hand.

—SHERLOCK HOLMES,
THE ADVENTURE OF THE ABBEY GRANGE

THE BAD NEWS—ADOLF HITLER BECAME CHANCELLOR OF GERMANY in January 1933. The good news—Prohibition ended in the United States in December 1933. The world was getting ready to go to hell, but at least Americans could face it with a drink in their hand.

For Percy Julian, the 1930s brought nothing but good news. After graduating from DePauw University in 1920 as class valedictorian, Julian earned a master's degree from Harvard in 1923 and then moved to Austria (with funding from the Rockefeller Foundation), where he obtained his PhD from the University of Vienna. His doctoral thesis focused on the medicinal chemistry of various plants. With his degree in hand, he returned to DePauw University in 1931 and embarked on a research career in chemistry.[1] The first project Percy Julian undertook in Indiana could not have been more challenging.

Julian's work in Vienna had involved the study of alkaloids—specifically, the identification of the soothing ingredient found in the Austrian shrub *Corydalis cava*.[2] Success with this project led him to envision a

59

bigger endeavor—the synthesis of an alkaloid called physostigmine. Why physostigmine? Ophthalmologists like Conan Doyle knew the answer—physostigmine was one of the few potential treatments for glaucoma. Extracts of the Calabar bean had been used since 1855 to cause constriction of the pupil (which improves the drainage of intraocular fluid, thus reducing pressure within the eye), and in 1877 Ludwig Laqueur administered Calabar extract to reduce the elevated intraocular pressure caused by glaucoma.[3] The active ingredient turned out to be physostigmine, a substance first isolated and identified in 1864 by Julius Jobst and Oswald Hesse. But extracting physostigmine from Calabar beans was an expensive, tedious process. A way was needed to synthesize physostigmine in a pure form, starting from simple, cheap precursors. But no one had ever synthesized a drug of this class—an alkaloid—before.

It's not that people weren't trying.[4] Sir Robert Robinson, world-famous chemist and international leader in the field of organic synthesis, published a paper in the *Journal of the Chemical Society* in 1932 claiming to have synthesized the critical alkaloid-like ingredient needed to make physostigmine. He was quickly challenged by another blue ribbon organic chemist, Toshio Hoshino of Japan, who in 1934 published an article in the *Proceedings of the Imperial Academy of Japan* claiming he was the first to produce the critical physostigmine substrate. The two international Goliaths were battling fiercely when Percy Julian stepped into the academic arena and claimed that he was the only one to have synthesized the precursor in question.

Julian was at an immediate disadvantage. He lacked the reputation of an acknowledged authority like Robinson, and his institution, DePauw, paled in the eyes of the chemistry community when compared to a place like Oxford. Nonetheless, Percy Julian felt confident enough to challenge the prevailing superstars with his claim that "he alone . . . had the correct synthesis" for the crucial physostigmine precursor.[5] Julian's challenge resulted in a typical scientific pillow fight, but when the feathers cleared the man from DePauw was left standing. Percy Julian had indeed achieved the first complete synthesis of an indole alkaloid—in this case, physostigmine. His triumph was published in 1935.[6]

It was a spectacular achievement for any chemist, but not enough to earn Julian the chairmanship of DePauw's Chemistry Department. Racism was likely a factor in this disappointing outcome. Pursuing other avenues for advancement, Julian took a position with the Glidden Company in Appleton, Wisconsin.[7] He became an expert in soybean products and developed pathways to synthesize numerous substances starting from soy-based ingredients.[8] One of his soy-related developments would revolutionize the cortisone story.

Once physostigmine could be synthesized from chemical precursors, it was no longer necessary to extract it from other sources (like the Calabar bean). But obtaining naturally occurring pharmaceutical agents by extracting them from appropriate sources was still the rule. As noted earlier, estrogens derived from human or horse urine were commercial success stories. If Kendall was correct, extraction of the adrenal cortex to obtain "cortin" would become just as successful. As the 1930s beckoned, Kendall set forth to extract, identify, and hopefully synthesize the mysterious substance found in the adrenal cortex.

By 1930 the Henches were beginning to collect assorted works and paraphernalia related to Sherlock Holmes. In some ways it was a curious hobby for Philip Hench to pursue—one eventually tinged with subtle irony.

The detective's brilliant deductive mind obviously appealed to the scientist in Hench. But how did he rationalize Holmes's negative attributes—for example, his cavalier drug use? And what of the paradoxes involving Holmes's creator, the late Sir Arthur Conan Doyle? Doyle had many characteristics that Hench would have found admirable—accomplished physician, fierce moralist, and superb writer. Would this admiration have been lessened by the dramatic personality transformation that overtook the famous author of the Sherlock Holmes stories later in life?[9]

Surely a Holmes scholar like Philip Hench knew of Doyle's moral eccentricities and intellectual faults. Was he likewise aware of the subtle connection between Sir Arthur Conan Doyle and the mutilation of cattle? In 1903 an Englishman named George Edalji was arrested and charged with entering a farm and mutilating cattle during the night.[10] Doyle, whose crime-solving abilities were sometimes confused in the public eye with those of his famous literary creation, was drawn by his reputation into the case. In these situations the writer's involvement was rarely productive. But in this particular instance Dr. Doyle was able to use his familiarity with the field of ophthalmology to demonstrate that Edalji could not have been responsible for the atrocities of which he was accused; the suspect had a form of night blindness, and his impaired nocturnal vision would have made it impossible for him to navigate the rugged terrain of the crime scene.[11] Doyle eventually published his conclusions, and the ensuing publicity led to a reopening of the case. Edalji was ultimately cleared of the crimes.

This anecdote involving Conan Doyle is noteworthy considering that cattle mutilation on a massive, unprecedented scale now enters the cortin story. In order to extract cortin from adrenal glands, one must first obtain adrenal glands. Where would Kendall find the tons of glands he'd need to perform his extractions? The slaughterhouse in St. Paul from which Szent-Györgyi had obtained the glands he used to make hexuronic acid was now defunct. What next? The solution to his supply problem involved a stroke of business genius that rivaled anything Kendall ever achieved scientifically.

Kendall contacted a few of the people he still knew at Parke-Davis and negotiated a novel proposal. He realized that the giant pharmaceutical company was already extracting a product from adrenal glands—chemically, it was known as epinephrine, but commercially in the United States it had been trademarked under the name Adrenaline. Parke-Davis was making a substantial profit from the extraction and sale of Adrenaline. But a "substantial profit" can always be made even more substantial, and Parke-Davis was looking to cut production costs. Taking a page out of today's global approach to business, Kendall offered to let Parke-Davis outsource the production of Adrenaline to his laboratory at the Mayo Clinic.[12] His offer: Parke-Davis would provide him with a steady supply of bovine adrenal glands. He would extract Adrenaline from the portion of the gland where it resides (specifically, the adrenal medulla) and turn it over to the company. Once this was completed, the remaining glandular tissue, including the cortex where cortin was found, would be his to use as he wished.

The company went for the idea. A deal was struck in which Kendall would be held to the same high purity and yield standards established for the Parke-Davis laboratories. As long as Parke-Davis received the same amount of pure Adrenaline from Mayo as it would have expected from its own facility, the company would provide him with 600 pounds of adrenal glands every week.

But Kendall was leaving nothing to chance. The extraction and purification of Adrenaline is a complex, difficult process: make the slightest mistake in any step along the way and–poof—everything you were working on is gone. Unsure of his ability to produce Adrenaline as efficiently as Parke-Davis, he struck a second deal with Wilson Laboratories, a subsidiary of the Wilson Packing Company of Chicago. At the time, Wilson Laboratories was making its own brand of adrenal cortical extract, a solution akin to that made by Pfiffner and Swingle. The potency of this extract varied tremendously between batches. Poor standardization made the extract difficult to administer clinically, which limited its commercial potential. Using Ingle's techniques, Kendall agreed to standardize the extracts for the Wilson Company in exchange for another 300 pounds of adrenal glands per week. The chemist could use these glands any way he wanted, including the extraction of epinephrine (Adrenaline), and he could keep any of the epinephrine he obtained.

Now Mayo had a buffer for the deal with Parke-Davis—if Kendall provided Parke-Davis with the epinephrine from the company's 600 pounds of adrenal glands per week, along with that extracted from the 300 pounds per week provided by Wilson Laboratories, there was no question that Mayo would supply enough Adrenaline to meet the yield standards required to keep Parke-Davis happy. But that meant processing 900 pounds of adrenal glands every week.

As Kendall tooled up his laboratory to handle the massive adrenal gland extraction operation, he once again displayed a little of that famous intuition of his—intuition that, unfortunately, was famous mostly for getting him into trouble. Kendall had already discovered that the adrenal cortex produced a significant amount of lactic acid, a common metabolic waste product. It occurred to him that cortin just might be a compound made by joining lactic acid manufactured in the cortex to epinephrine produced in the medulla. "Perhaps," he thought, "cortin is lactyl epinephrine?" Never fearing to "seek a complex explanation" when a simpler one might suffice, Kendall set to work looking for lactyl epinephrine in his adrenal extracts.[13]

To no one's surprise, Kendall quickly announced that he had identified the presence of lactyl epinephrine in his adrenal extracts. (Most of his coworkers, however, doubted that he had really found it. They believed he had isolated both lactic acid and epinephrine from the adrenal gland, but nobody believed the two chemicals were actually joined together into a single molecular entity.)[14] Kendall reported at a Mayo Clinic staff meeting that "these observations suggest that the derivative of epinephrine present . . . may be lactyl epinephrine." His remarks that evening were later published in the *Proceedings of the Mayo Clinic* of March 2, 1932.[15]

Kendall subsequently tried to synthesize lactyl epinephrine, and for a short period of time he thought he was successful in doing so (according to his coworkers, it's not clear what, if anything, he actually synthesized). When tested on adrenalectomized animals, the lactyl epinephrine made by Kendall had minimal, if any, physiological effect. It quickly became obvious that this was not the "cortin" scientists were looking for. When he finally became convinced that lactyl epinephrine wasn't cortin, Kendall returned once again to his hunt for the elusive substance.[16]

An operation was established in Kendall's Mayo Clinic laboratory for the commercial extraction of Adrenaline, as required by the deal with

Parke-Davis. This adrenal-processing "factory" operated twenty-four hours a day in three shifts. It ran like a factory, and it even sounded like a factory. According to Kendall, "The constant rushing sound from the six water-suction pumps plus the refrigeration unit imparted a sense of power and urgency."[17] It must have been impressive. Between 1934 and 1939 the Kendall lab processed 150 tons of adrenal glands and produced Adrenaline worth, at that time, over $9 million.

Kendall was forever marveling over the subtle mysteries of his own work. After extracting Adrenaline from the medulla of the glands, he began searching through the crushed residue for cortin. The pungent, raw adrenal tartare was dissolved in water, alcohols, acids, bases, benzene, and myriad other solvents that were later boiled, frozen, evaporated, or mixed with different agents. The various extraction techniques eventually began to produce trace amounts of the precious substances hidden within the mushy brown tissue. Kendall considered this miraculous. "To start with material that resembles hamburger and to separate a few milligrams of crystalline material is similar to bringing a pearl of great beauty from the murky depths of the sea."[18]

Just as his colleagues were beginning to forget the lactyl epinephrine debacle, Kendall returned to the hot seat. In 1933 he once again announced at a weekly meeting of the Mayo Clinic staff that he had finally isolated the elusive "cortin" for which he and other scientists had been searching. An account of this seemingly historic evening was later published in the *Proceedings of the Mayo Clinic* on April 25, 1934.[19] According to Kendall, the substance he had identified was a single entity for which he had derived a formula; he also implied that he'd established the chemical structure of the molecule. He waffled somewhat on the potency of his new compound, pointing out that there were still unresolved issues to be dealt with regarding the standardization of the preparations. But in some areas his claims were unequivocal. For example, he offered that "in this instance, it may be said that administration of the crystalline product will maintain a dog in normal condition after double super-adrenalectomy."[20] The ability to maintain the life of an animal after removing both of its adrenal glands is the ultimate test for cortin, and if Kendall's discovery could do this, he had indeed found the elusive substance.

By the end of his presentation it was clear that the clinic's leader, Dr. William Mayo, was willing to forgive Kendall's earlier mistakes regarding lactyl epinephrine. The steely-eyed chief physician and head of the clinic stood, applauded, and began to heap accolades on his chemistry colleague.[21]

If this was the beginning of the end of the cortisone story, the timing couldn't have been more perfect, because at that exact moment a young man from the East Coast was visiting the Mayo Clinic as a patient. He had been diagnosed by his home physicians as having colitis, and although cortisone would ultimately prove extremely useful in the treatment of this condition, no good therapy for colitis was available at the time. The seventeen-year-old was not exactly happy to be in Rochester, or any place for that matter, where neither a cure nor an effective treatment was available. Sounding like the frustrated teenager he was, he wrote from Rochester to his friends back home that he was stuck in "the God-damnest hole I've ever seen."[22]

The young man was John F. Kennedy.

Stories of JFK's health problems are a matter of public record and have been widely reported by others. According to many published accounts, Kennedy was a somewhat sickly child, and during his junior and senior years of prep school at Choate, his blood count had to be carefully monitored; there were concerns that he might be developing leukemia or agranulocytosis (a decrease in certain white blood cells). Jack's parents sent him to the Mayo Clinic, where he spent "a miserable month." Alone in Rochester, he struggled the first two weeks as an outpatient at the clinic before being admitted to Saint Marys Hospital for additional tests and treatment.[23]

There was no treatment—yet—for the problems that ailed young JFK. But that was about to change. And despite his uncomplimentary rhetoric about the clinic—remarks certainly made out of bravado and frustration rather than serious criticisms about his care—he'd eventually be back.

CHAPTER 8

Kendall Strikes Out Again

It is a capital mistake to theorize before one has data. Insensibly one begins to twist facts to suit theories, instead of theories to suit facts.

—SHERLOCK HOLMES, "A SCANDAL IN BOHEMIA"

BY 1931 THE GREAT DEPRESSION WAS FINALLY EASING, BUT NOW Hemingway was getting vexed. His latest manuscript, which would eventually be called *Death in the Afternoon*, was coming along well enough. But his friendship with F. Scott Fitzgerald was not.

Ernest Hemingway and F. Scott Fitzgerald had met for the first time at the Dingo Bar in Paris in April 1925, shortly after the publication of *The Great Gatsby*. They initially became friends—drinking, exchanging advice, and offering support for each other's careers. Their friendship later turned uncomfortable.[1] Fitzgerald's wife, Zelda, disliked Hemingway immediately, describing him in very negative terms and suggesting that his macho quality was "as phony as a rubber check."[2] She became delusionally convinced that Hemingway and her husband were having an affair, which if true would have certainly been the antithesis of their testosterone-driven images. Zelda Fitzgerald and Ernest Hemingway turned out to have more in common than the handsome Mr. Fitzgerald—they would ultimately share mental illness, repeated institutionalizations, and various forms of antipsychotic convulsive therapy.[3]

By 1931 several key scientific players in the unfolding cortin saga—Russell Marker, Philip Hench, Edward Kendall, and Albert Szent-Györgyi—were, like Ernest and Zelda, moving along separate paths that would eventually converge in unpredictable ways.

Russell Marker, now established at Pennsylvania State University, was just beginning his work on steroids. He was particularly interested in progesterone, which was not only an important hormone in its own right but also a key ingredient in the manufacture of many other steroids. Parke-Davis (which provided the adrenal glands for Kendall's work) was helping Marker in his efforts by supplying him with the necessary precursors of progesterone, most of which involved derivatives made from horse urine.

Philip Hench was in hot pursuit of new treatments for rheumatoid arthritis. He monitored reports of every novel therapy purported to offer a chance of helping the disease: fever therapy (typically produced by typhoid vaccine or bee stings); radiotherapy; injections of gold, olive oil, or histamine; dietary changes; vitamin therapy; bone marrow stimulation; and even tonsillectomy. One treatment in particular caught the Mayo rheumatologist's interest: the use of a liver toxin called cinchophen.[4] Its administration seemed to help some patients, but not others. The mechanism of action was not known, but it produced jaundice in many patients. Philip Hench couldn't stop thinking about the relationship he had observed in the past between jaundice and the relief of rheumatoid arthritis. He continued to collect and report on additional patients with jaundice-induced remission of rheumatoid arthritis.

Edward Kendall was busy building the required laboratory infrastructure for this type of research. His most pressing problem? If he wanted to make a serious foray into the search for cortin, Kendall realized he needed the capacity to measure cortin's biological activity. This was an area of physiological expertise he did not possess, and for once the chemist seemed to understand that he was only as good as the people with whom he worked. He solved his problem by hiring Dwight Ingle in 1934. According to Kendall, Ingle was "endowed with an insatiable urge to find out why certain things are so. One of these things was . . . the endocrine glands."[5] Ingle had been trained in psychology at the University of Idaho, and later at the University of Minnesota, but he aggressively pursued studies in physiology after becoming convinced that psychological diseases were really manifestations of physiological problems. In an attempt to better understand the actions of certain newly discovered hormones, Ingle developed his own unique techniques and devices for assessing the physiological activity of these compounds.

One apparatus Ingle invented required the use of an anesthetized rat; the rat's leg muscles were electrically stimulated, causing them to twitch, and the strength of the twitch was measured. With this device the ability of various drugs to affect muscle function could be tested.[6] Previous work had shown that removal of the rat's adrenal glands caused the electrically stimulated muscles to cease twitching in less than twenty-four hours.[7] But when cortin-containing solutions, such as those produced by Pfiffner and Swingle, were given to the adrenalectomized rat, the muscle twitch response could be maintained indefinitely. The more potent the cortin-containing solution, the better the response.

Ingle's experimental expertise and overall scientific knowledge gave Kendall a significant edge over his biochemistry rivals. Ingle brought one other quality to the lab as well: a sense of scientific order. Ingle's mantra could have been summed up by a line from Sherlock Holmes's *The Adventure of the Retired Colourman*: "Things must be done decently and in order."[8] As noted earlier, Ingle had concerns from the outset about Kendall's scientific savvy and discipline. He reported that "Dr. Kendall neither tested ideas in private nor concealed his mistakes, as many of us do. He would announce what he expected to find 'just around the corner.'" Ingle recognized the need to counterbalance Kendall's blind enthusiasm. It seems that Ingle, with his meticulous attention to the details of scientific process, would join Nick Kendall's laboratory just in the nick of time.

Albert Szent-Györgyi wasn't necessarily a "key player" in the cortin story, but he was nevertheless playing an interesting peripheral role at this time. After returning to Europe, he set up his own laboratory in Szeged, Hungary. In late 1931 he acquired the assistance of an American of Hungarian descent, Joseph Svirbely, who had trained with the prominent American chemist Charles G. King at the University of Pittsburgh. King and Svirbely had been attempting to isolate vitamin C; they seemed to be closing in on success when Svirbely abruptly left, went to Hungary, and joined Szent-Györgyi's group.

Svirbely's route into the Szent-Györgyi lab was nontraditional. He simply appeared unannounced in Szent-Györgyi's office one day, explaining that he was a recent PhD graduate from America and that he was looking for work. When asked what special skills he possessed, Svirbely replied, "I can tell you whether something contains vitamin C or not."[9] Taking him up on his offer, Szent-Györgyi provided the young man with some of the remaining hexuronic acid that he had extracted at the Mayo Clinic. Svirbely, using an approach he had perfected in America, tested it on guinea pigs with artificially induced scurvy (a disease caused by a lack of vitamin C). Amazingly, the animals given hexuronic acid recovered, proving that hexuronic acid was indeed the "vitamin C" that King and others were frantically racing to isolate and identify. In retrospect, Szent-Györgyi had always suspected that hexuronic acid was vitamin C—even after preliminary studies conducted by a friend in another laboratory had erroneously suggested it wasn't.

Svirbely wrote to King, his previous mentor, in early 1932 and told him what he and Szent-Györgyi had discovered;[10] he added that they were submitting a report to the prestigious journal *Nature.* A month later, on April Fool's Day, 1932, the equally prestigious journal *Science* published an announcement by King that he had discovered vitamin C, and that it was identical to the hexuronic acid isolated by Szent-Györgyi earlier. A "bitter academic controversy" ensued.

The team of Szent-Györgyi and Svirbely claimed the discovery as their own.[11] King, who was widely known to have been working on the issue of vitamin C isolation for more than five years, held his ground. In a miniature reenactment of the War of 1812, the American scientific community rallied around King, while the Europeans supported their countrymen. In the end, Szent-Györgyi's work was officially recognized as the first identification of vitamin C.[12]

Szent-Györgyi proved to be an astute politician as well as a clever scientist. Rather than trying to patent and control the production of vitamin C, he sent samples to every researcher in the world who was working on problems or on issues related to vitamin C (which he had renamed "a-scorbic" acid because of its ability to prevent scorbutus, more commonly known as scurvy). Supplies were sent via the League of Nations to countries like Norway, where scurvy was still a public health problem. His respect in the scientific community soared—or at least it did until he, like Kendall, stumbled by making premature claims about his discoveries.[13]

Szent-Györgyi tripped himself up by becoming a vocal advocate for vitamin C, touting its potential value as a preventative or treatment for a variety of illnesses, including the common cold (foreshadowing Linus Pauling, the Nobel Prize winner who would likewise damage his reputation by over-hyping vitamin C half a century later). The Hungarian had great difficulty getting others to share his promotional interest in the compound, and once studies began to suggest that vitamin C was not the "miracle cure" he had dreamed it to be, Szent-Györgyi reluctantly returned to his research activities in other areas.

Back in Rochester, this should have been Kendall's finest hour. After all, his amazing discovery of "cortin," which he had announced earlier at a Mayo Clinic staff meeting and for which he had drawn the accolades of Dr. Will Mayo, was still the talk of the clinic. Except for one little problem. Kendall was wrong. Again.

Most glands make and secrete just one primary substance. But Kendall, with the help of Ingle, was slowly coming to the painful conclusion that the adrenal cortex manufactures many different compounds. Unfortunately, some of these—including the substance he had announced at the Mayo meeting as being *the* cortin—couldn't sustain muscle activity in Ingle's rat assay and were unable to keep adrenalectomized dogs from succumbing to Addison's disease.[14] Now that Kendall had erroneously concluded he'd identified cortin—and publicly announced its discovery to his colleagues—he faced the kind of embarrassing situation that would have ended the careers of many scientists.

They say that timing is everything. That goes double for bad timing. Dr. Frank C. Mann was an animal surgeon at Mayo who prepared many of the experimental dogs used in Kendall's studies. At the time Kendall made his biggest mistake ever, Mann was—unfortunately for Kendall—a member of Mayo Clinic's ruling body, the board of governors. According to Ingle, "Mann was well aware of Kendall's foibles as a scientist," and he became "deeply concerned" about the overall conduct of Kendall's research.[15] Specifically, Mann was bothered by the claims Kendall had made previously about the "cortin" effects of lactyl epinephrine, which were stated in public but for which no significant publication was ever issued to back them up. He was now even more concerned about Kendall's recent claim to have isolated "the hormone of the adrenal gland."

Mann brought his concerns regarding Kendall's performance to the other members of the clinic's board of governors and aired them. After considerable debate, the board decided that another chemist in Kendall's lab, Harold Mason, should be "given increased responsibility for purification and chemical characterization."[16] It was a slap in the face for Kendall, and perhaps not too surprisingly he dragged his feet at implementing the board's "recommendation." When it became apparent to everyone in his laboratory that Kendall was not distributing responsibility as institutionally instructed, the team faced another dilemma. Should they complain again to Mann and the board of governors, or attempt to work this out face-to-face with Kendall? Mason, along with Charlie Meyers (another chemist working in Kendall's lab on the adrenal cortex extractions), picked the second option and met privately with their lab chief. Kendall capitulated; he reluctantly agreed to a series of weekly meetings in which he would accept input from other members of the team. The quality of the laboratory's work improved immediately. Kendall's only public comment regarding any of this appears to be that "Dr. Mason assumed the heavy responsibility of micro analyst . . . the results . . . were always reliable and the results progressed smoothly."[17]

It's hard to believe that Kendall was unaware of the controversies surrounding his abilities; more likely he recognized them but refused to acknowledge his coworkers' perceptions or actions. But whether he was clueless or merely indifferent, Kendall went back to his work as if success were just around the corner. As Ingle put it, Kendall "was a quiet man but did not conceal his enthusiasm."[18]

Clueless? Indifferent? These are harsh criticisms of Kendall—and perhaps not entirely fair. After all, the United States had just ridden out a severe economic depression during which Kendall needed to keep reselling his costly projects to the Mayo board of governors every year. Lesser men might have let the clinic close down their lab (as it did to Rowntree and Alvarez), which would have been the end of this story. No lab, no discovery of cortisone. In retrospect, Kendall did what he needed to do in order to keep the lab open and his projects on track. If that meant generating excitement and enthusiasm by pursuing questionable leads, jumping to conclusions, making premature announcements, grossly exaggerating the truth, or doing any of the other cliché "bad behaviors" cited by others and used to disparage him, these things were, nonetheless, key to his ultimate success.

Kendall had, after all, once gold-plated a bronze medal for rowing—but only because he knew he'd deserved a gold one. Surely gold-plating some equally worthy bronze research projects could, under the right circumstances, be justified.

But there was more bad news on the horizon for Kendall. In Switzerland, at a place not far from Reichenbach Falls—the spot where Professor Moriarty killed Sherlock Holmes—Tadeus Reichstein was preparing to undertake his own quest for cortin. The implications of this 1934 event to the cortisone story cannot be underestimated. While it remains debatable whether Ingle was right about Kendall when he said he was "not a great chemist," there was never any doubt about Reichstein's talent with a test tube. The Swiss genius was the Jesse Owens and Roger Bannister of chemistry rolled into one. Kendall, in contrast, was a slow plodder.

Nick Kendall now faced a serious problem. A much faster thoroughbred was entering the cortin race.

CHAPTER 9

Kendall Presses On

I confess that it is not the situation which
I should like to see a sister of mine apply for.

—SHERLOCK HOLMES,
"THE ADVENTURE OF THE COPPER BEECHES"

NINETEEN THIRTY-SIX WAS NOT LOOKING LIKE A GOOD YEAR FOR THE ladies.

Amelia Earhart was missing over the Pacific.

Geraldine Smithwick, a senior at the University of Chicago, thought she was having a good year, but it turned out she was wrong—she just didn't know it yet. She'd recently married Luis Alvarez; the young physicist from Rochester had passed his oral exams and received his PhD a few days earlier. In a month they would be heading off to his new job in Berkeley, California. But soon the war would separate the beautiful socialite from the scientist; they'd grow apart and eventually get a divorce.

Mabel Alvarez, Luis's aunt, was also in trouble. Artistically, things were never better. But her love life was delusional.[1]

That love life centered on Dr. Robert H. Kennicott, a Los Angeles physician who had been born in Luverne, Minnesota, in 1892 (a southern Minnesota town much like Rochester) and moved to Los Angeles in the

1930s. Mabel met him through her brother, Dr. Walter Alvarez, who had introduced the pair to each other before he moved from California to Mayo. Kennicott was a prominent physician in Beverly Hills with a celebrity-oriented private practice that was unrivaled in Tinseltown.[2] But celebrity medicine wasn't driving the initial attraction between Mabel Alvarez and Robert Kennicott. It was art. Dr. Kennicott was more than a high-profile physician to the stars; he was also an accomplished artist, a man with a talent for drawing and painting the human figure. He and Mabel attended art events throughout the Southern California art community—exhibitions, sketching sessions (in which they often shared the same models), galas, gallery openings, parties, or whatever the social happening du jour offered.

The two bohemians were "a natural match" for each other. Mabel, by far the better artist, encouraged and influenced Kennicott's artistic style and direction, but her interest in him went well beyond art. She was in love with the handsome, charming doctor, and the romance for which she longed would span nearly a decade. Her diaries described many occasions when she was assumed to be "Mrs. Kennicott" at various social events; she blissfully fantasized about that possibility and the idea of living "happily ever after" with her dream man.

But love's existential voyage from Hollywoodland to Matrimonyland was doomed. As the years passed, there was frustratingly little forward progress in the relationship. Mabel Alvarez was blind to the reason for this, although nearly everyone else around her understood the cause with crystal clarity. Robert Kennicott "was more interested, it would seem, in the male models that he painted" than he was in Mabel. In 1939 she suddenly realized that she had "hitched her dreams" of happiness to "a gay man who would never give her marriage."[3]

In the end, Mabel's dreams of marriage were brushed over with the pigments of pain and humiliation like a flea market painting done on a piece of cheap black velvet. The eventual breakup was devastating for Walter Alvarez's little sister. "My life [is] aimless now," she wrote. "It . . . needs more context."[4] Moving to Hawaii, her childhood home, she escaped the embarrassment and pressure of Los Angeles. Once resettled, Mabel focused on rebuilding her art career. And, in the process, rebuilding her life.

Mabel Alvarez's delusions of love nearly destroyed her. But Nick Kendall's delusions were finally starting to pay off.

In fact, his search for cortin was becoming less delusional every day. Having barely survived his humiliating faux pas with the premature, misleading announcement that he had "found it," Kendall was back at work looking for cortin—and trying to put a good PR spin on the results he was getting. By 1936 Kendall had identified five separate compounds produced by the adrenal cortex.[5] Using the letters of the alphabet, he labeled them A through E based on their order of identification.[6] A sixth compound, which would be called F, was on the verge of being isolated and crystallized. All of these substances seemed to be chemically related; more surprising, all were conclusively shown to be steroids—members of a family of complex molecules constructed from interlocking carbon rings. It now seemed inevitable that cortin would turn out to be a steroid and therefore related to other steroid hormones like estrogen, progesterone, and testosterone.

To the surprise of many, Kendall was now taking a much more cautious, meticulous approach to his work. He made no apologies to his peers at the Mayo Clinic for his prior erroneous assertion that the substance he had isolated was *the* cortin. For Kendall, discovering that the adrenal gland made multiple steroid substances was sufficient redemption for his mistake. After all, he hadn't been entirely wrong with his original announcement—he just hadn't been entirely right. That, he believed, was simply the nature of scientific research. The possibility that he might have been fired from the clinic for his sloppy, error-riddled work seems never to have crossed Kendall's mind.

It was a good thing for Kendall that he had found these five steroid compounds, because Tadeus Reichstein in Switzerland had just isolated *seven* similar compounds from the adrenal cortex.[7] Pfiffner (at Princeton) gave up studying adrenal extracts a year earlier, but his partner, Swingle, moved to Columbia University (Kendall's alma mater) and joined the team of another very strong chemist named Oskar Wintersteiner. With Swingle's help, the Columbia group quickly isolated five of its own adrenal gland steroids.

Nomenclature was becoming a big problem. Each of the three major laboratories searching for cortin used a letter of the alphabet to designate its discoveries; the letter was usually based on the order in which the compounds were discovered in each lab. If different laboratories isolated compounds in a different order, the names didn't match up between the various labs and publications. For example, Pfiffner had a compound "F" that turned out to be the same as Kendall's compound "E," while the identical compound was called "substance Fa" in Reichstein's lab.[8] In the interest of science the three laboratories shared samples with each other for comparison purposes, but there remained a serious competition between the principals.

An apparent breakthrough came in 1937 when Tadeus Reichstein identified yet another entity in the adrenal cortex, which he named "substance H." Testing by the Organon Company (his source of raw materials for this work) in Holland had already demonstrated that substance H was the most active "cortin" he'd discovered so far. Its physiological effect on adrenalectomized animals was so impressive that even the quiet, conservative Reichstein finally claimed that he had identified *the* cortin, and he boldly gave it a name. Corticosterone. Its isolation had required the kind of exceedingly tedious work for which the Swiss are famous; in this case, 1,000 kilograms of adrenal glands (obtained from nearly 20,000 beef cattle) were processed into a few grams of the new material.[9]

The good news—corticosterone was an exciting, potentially huge discovery. The bad news—producing it was commercially unfeasible. But back at Penn State, Russell Marker was making excellent progress with research that would change this situation.

A year earlier, in 1936, the Parke-Davis company had sent Marker an extract made from everybody's favorite source of hormones, fermented pregnant mare's urine. Marker had identified a steroid in the extract, and from it he was able to isolate pregnanediol.[10] Using conventional chemistry methods, Marker converted this precursor into 35 grams of progesterone, which at that time was the largest batch of progesterone ever manufactured. Gold is worth about $45 per gram today. Back then, progesterone was going for $80 per gram. Valuable stuff.

At this point neither Kendall nor Marker nor Reichstein had any inkling that the three men's research projects were chemically related. What could extracts from the adrenal gland possibly have in common with decomposing fertility steroids present in rancid horse pee? The logical answer seemed to be "nothing." But this answer was wrong.

The discovery of corticosterone by Reichstein's group was perceived as a major blow to the Mayo Clinic's effort to find "cortin." As the implications of Reichstein's announcement became clear, a collective thought settled over Kendall's lab, the Mayo biochemistry group, and the clinic's board of governors. Was it finally time to quit the race to discover cortin?[11]

Kendall was surely sweating hollow-point .38s. He must have known that Reichstein was a better chemist and would eventually identify every nasty little secret the adrenal glands held. Kendall didn't have the intellectual ammunition to win a gunfight against Reichstein. In all likelihood the Swiss chemist was going to isolate and manufacture cortin before he did—if he hadn't done so already. On the other hand, Kendall knew he had some distinct practical advantages over his European competitor.

First, Kendall had more of the vital raw materials. Between Parke-Davis and Wilson Laboratories, Kendall was receiving 900 pounds per week of fresh adrenal glands. In contrast, Reichstein was getting small—and extremely expensive—shipments of Dutch adrenal extract from the Organon Company. Kendall had also heard some inside information suggesting that the executives at Organon were not exactly thrilled with the scientific payoff that Reichstein had returned to them so far. True, he'd made significant accomplishments, but they had expected much more. Reichstein might be an "in-law" member of the Dutch Royal Family, but that didn't mean the essential extract from Holland couldn't—or wouldn't—stop coming someday.

Second, Kendall had better physiological assays to work with, thanks to Ingle and his rat muscle stimulation apparatus. This device, along with Kendall's ability to utilize surgically modified animals (like the adrenalectomized dogs and rats that Dr. Mann provided), gave Kendall the ability to test his new compounds in ways that would always be off limits to Reichstein.

Third, and perhaps most important, Kendall had the Mayo Clinic behind him. The Mayo Clinic had doctors. And patients. This combination gave Kendall the ability to do clinical trials on humans with his new compounds, which was something that Reichstein could only fantasize about. When it comes to conducting medical research on patients, there's no substitute for a hospital like the one Sister Alfred and Dr. W. W. Mayo had built.

It's unclear whether these arguments would have been enough to justify continuation of the cortin project. However, two additional pieces of good news arrived while the decisions were being pondered. First, additional testing was performed on Reichstein's substance H, and it was determined that the biological activity of this "cortin" was much less than originally thought. True, it had a big effect on the ability of muscles to perform work (as noted in the rat muscle twitch device), but it was relatively ineffective at correcting the salt and water imbalances that occurred in animals following adrenalectomy. If it couldn't correct this aspect of Addison's disease, it must not be *the* cortin. The second piece of good news came as a total surprise. Upon further testing, it was clear that Reichstein's new substance H was identical to Kendall's old "compound B," which he had isolated more than two years earlier.[12]

Even if substance H turned out to be the cortin for which they were looking, Kendall had found and claimed it first.

The Mayo Foundation leadership now faced a difficult decision. Kendall's work held great potential. But his discoveries to date were of uncertain significance. It was possible that he would never identify *the* cortin—and if he did, cortin could still turn out to be far less valuable than everyone believed it to be. Was it time for the clinic to cut its losses and put an end to Kendall's expensive, questionable adventure?

The answer? Over forty years earlier Sister Alfred approached Will and Charlie's father and proposed building a hospital in their tornado-devastated town. It seemed like a long-shot proposition—she had no nurses, no money, and no experience in running a medical facility. W. W. Mayo was naturally reluctant to proceed. But once Sister Alfred got him to commit to the endeavor, everyone knew he was going to stick with it. Dr. Mayo was a man of his word, and he knew that in a small town like Rochester every project is a major project. For something as costly and important as the building of a hospital, the consequences of failure would be financially and socially catastrophic for the community. While everyone accepted that construction setbacks would occur, there was an unspoken rule that the participants would forge on despite them. And eventually succeed. They would not be stopped by cost, politics, or their own inexperience. Perseverance would overcome all.

This 1880s frontier attitude still remained at the heart of the clinic's 1930s philosophy. The Mayo brothers, and the foundation's board of governors, were in agreement—despite the earlier scientific and social mistakes made by Kendall, the relatively slow pace of his progress, his lack of definite success, and the tremendous cost his experiments were incurring, the clinic would continue to support him and his research.

CHAPTER 10

Score: Szent-Györgyi–1; Kendall–0

I can't make bricks without clay.

—SHERLOCK HOLMES,
"THE ADVENTURE OF THE COPPER BEECHES"

WINNING THE NOBEL PRIZE TRIGGERS AN EMOTIONAL, VISCERAL response, and for Albert Szent-Györgyi, the October 1937 announcement that Sweden's Karolinska Institute was bestowing the Nobel Prize for Physiology or Medicine on him was truly overwhelming. The award, honoring "his discoveries in connection with the biological combustion processes, with special reference to vitamin C and the catalysis of fumaric acid,"[1] sent a wave of Hungarian pride across the nation. When the Nobel committee called to inform him of its decision, Mrs. Szent-Györgyi was at home and took the message. She made sure that most of the city of Szeged knew who won the award before Szent-Györgyi himself found out.

Across the ocean, Nick Kendall was taking the announcement in stride. Szent-Györgyi was a friend and colleague, and Kendall surely wished him well for his good fortune. But with old memories of the thyroxine debacle still lingering, the Mayo Clinic chemist couldn't help feeling that he and his lab had once again teetered on the brink of greatness only to come away empty-handed. After all, Szent-Györgyi had worked under Kendall in Rochester, had isolated large amounts of hexuronic acid (now referred to as vitamin C) in Kendall's lab, had left the raw material with Kendall, and had offered to let Kendall become an intricate part of this project. It was Kendall who opted out, apparently believing that the further study of hexuronic acid wasn't worth the potential payoff. In retrospect, the Mayo chemist had made the wrong professional choice. Again.

In the building across the street from Kendall's lab, Dr. Walter Alvarez was dealing with his own difficult issues. Alvarez had found a unique clinical niche for his medical practice: he was "collecting" patients with odd or unexplained illnesses, especially those with gastrointestinal problems. And John Kennedy, a young man with an exceptionally challenging digestive tract, was once again back at the clinic.[2] The once-feared possibility that JFK had some type of low-grade leukemia seemed less likely now. As a teen, his white blood cells had fallen to uncomfortably low levels; in 1935 Kennedy had written from the Peter Bent Brigham Hospital in Boston that his white blood cell count had dropped from 6,000 to 3,500 over a three-week period. "At 1500 you die," he told his friend. "They call me '2000 to go Kennedy.'"[3] But while his blood was no longer thought to harbor a serious problem, his bowels remained a problem. He was losing weight, passing bloody stools, and suffering considerable abdominal distress.[4]

Another trip to Mayo might seem a little less onerous if it led to relief, but unfortunately JFK may have suffered from ulcerative colitis (in which the colon becomes inflamed, ulcerated, and bleeds), a condition for which there was (and still is) no medical cure, and for which the conventional treatments of the time brought only minimal success.[5] Based on published accounts in the January 1934 and December 1936 issues of the *Mayo Clinic Proceedings*, the usual therapy for colitis involved a combination of restricted diet, reduction in emotional stress (thought to be a major contributor to both the colitis and the ulcerations), and injection of serum obtained from horses.[6] The doctors at Mayo must have believed there was some utility to this treatment regimen (after all, it had been recently published in the clinic's medical journal), but they surely knew it was no cure.

However, in 1937 there emerged a potential treatment for colitis that seemed promising. This glimmer of hope came in the form of a steroid that had been previously isolated from the adrenal gland. Desoxycorticosterone acetate (a close relative of Reichstein's corticosterone, or Kendall's compound B,) often abbreviated as DOCA, had been tested in animals, where its main effect seemed to be on the body's absorption and secretion of salt— but it also possessed a limited ability to suppress inflammation. When given to animals or patients with Addison's disease, it generally helped them live longer. Unfortunately, DOCA by itself couldn't restore patients with adrenal insufficiency to normal health. Indeed, some patients with Addison's disease who were treated with DOCA actually died from side effects caused by increased fluid retention and decreased potassium levels. Other drug-related complications such as paralysis[7] and life-threatening heartbeat irregularities[8] were common. But despite its imperfections, DOCA was still the closest thing to "cortin" that had been discovered so far, and there was keen interest in using it on patients with Addison's disease. This interest was heightened when Reichstein's group in Switzerland found a practical way to synthesize DOCA from bile acid, making it commercially available. And outrageously expensive.

Kennedy historian Robert Dallek writes, "it is now well-established that Kennedy was treated with DOCA . . . it is possible that Jack was taking DOCA as early as 1937."[9] Dallek describes an episode in which Kennedy, in a handwritten note to his father, laments over the difficulties involved with getting a prescription filled in Cambridge, Massachusetts, where he was attending Harvard. Kennedy surely realized that his medicine—whatever it was—was rare and controversial. Unable to obtain the drug in any local pharmacy, he wrote to his father and asked him to try to find a source: "I would be sure you get the prescription," he urged, adding, "some of that stuff . . . is very potent and [my doctor] seems to be keeping it pretty quiet."[10] Years later one of Kennedy's friends would describe him implanting a "pellet" in his leg, using a little knife to cut the surface of the skin and slip the pellet underneath before putting a bandage over the top of it.[11] This was the usual way in which DOCA (a poorly absorbed drug when taken orally) was administered at that time—direct implantation under the skin permitted proper absorption. But was Kennedy really taking DOCA? Or had he been prescribed some other new medicine?[12]

What investigators could not have possibly realized in the 1930s was that many patients would develop serious side effects from using DOCA or other steroids. Indeed, John Fitzgerald Kennedy was destined to become, in due time, the poster child for adverse steroid-induced consequences.

Walter Alvarez wasn't the only doctor collecting patients with oddball diseases. Philip Hench was doing the same. As noted earlier, in 1929 Hench observed that certain patients with rheumatoid arthritis went into remission after developing jaundice or becoming pregnant. He was now "collecting" other patients with similar remissions. Hench was also designing experiments in which he planned to artificially induce jaundice in patients with rheumatoid arthritis, after which he'd determine whether the jaundice produced a beneficial response on their symptoms or disease. One thing was certain—Hench's scientific work would surely be meticulous in its conduct, precisely executed, and impartially analyzed. It was already widely recognized around the clinic that high standards of caution and precision were basic attributes of Dr. Hench.

At roughly this same time—and in utter contrast with his scientific interests—Hench also became interested in the (legendary) death of Sherlock Holmes at Reichenbach Falls near Meiringen, Switzerland. He collected newspaper articles, magazine clippings, travel pamphlets, and countless descriptions of the falls and surrounding areas. His friends, watching him tackle such a mundane problem with so much scientific precision, couldn't help but recognize that Hench was just as Holmes-like in the conduct of his daily activities as he was in the pursuit of his medical research.[13] Indeed, Hench's highly choreographed, meticulous, conservative approach to science and research was as different from Kendall's style—reckless optimism, constant improvisation, and "fly-by-the-seat-of-the-pants" tactics—as humanly possible.

Hench was also discovering a new interest, something that combined the medical mystery of "substance X" with the deductive reasoning and super-sleuthing abilities of Sherlock Holmes. The topic? Dr. Walter Reed and the story behind the conquest of yellow fever.

Hench's close friend Ralph Cooper Hutchison, president of Washington and Jefferson College, was preparing to dedicate the college's new chemistry building in honor of one of the college's most famous graduates, Jessie W. Lazear, an alumnus who had died in Cuba while working on Reed's yellow fever experiments. With Hutchison's approval, Hench began to contact some of the other original fever research participants. Hench would soon discover that much of the official "history" regarding Reed's work was inaccurate or simply wrong.[14] It seemed that egos; conflicting descriptions of events; wild, unverified claims; and partisan interest between the governments of the United States and Cuba were clouding the "interpretation of the facts." Some of those who had participated in Reed's original work urged Hench to "clear up" the story. Hench became fascinated by the army doctor and his famous research, and he eventually agreed to "document every aspect" of Reed's work and write the ultimate "accurate and comprehensive history."[15] Given Hench's nearly pathological attention to detail and precision, it should surprise no one that he spent much of the next two decades pursuing this "hobby."

While Szent-Györgyi basked in the glory of the Nobel Prize, and Alvarez and Hench trudged through the trenches of clinical practice with their odd-ball patients, Russell Marker was, in the words of Sherlock Holmes, trying to make bricks without clay. Or, less figuratively, he was trying to make progesterone without the proper ingredients. The urine-derived starting material he had received from Parke-Davis worked fine, but it was expensive and difficult to obtain. An alternative building block was needed. And Marker had an idea.

His brainstorm involved the use of sarsasapogenin, a steroid derived from the sarsaparilla plant. The conventional chemical wisdom of the time held that sarsasapogenin, which was chemically similar to progesterone, could not be turned into progesterone because sarsasapogenin was chemically inert. Marker, in a brilliant burst of chemistry insight, hypothesized that sarsasapogenin was not only chemically reactive, but that he could convert it to progesterone. He devised a series of reactions that used sarsasapogenin as a starting product, and as predicted, the final result was progesterone. This chemical process became known as the Marker degradation; it remains one of the true landmarks in the history of steroid synthesis.[16]

Marker thus succeeded in making progesterone without the usual precursors. But a problem remained. Sarsasapogenin was almost as rare and expensive as the urinary precursors he'd been forced to use previously. It was no more economically feasible than extracting progesterone from concentrated urine. But by using the Marker degradation it was now possible to degrade other plant steroids to progesterone. Marker began searching for a cheaper, plant-derived chemical cousin of sarsasapogenin from which he could synthesize progesterone.

CHAPTER 11

Transitions and Travels

Work is the best antidote to sorrow.

—SHERLOCK HOLMES,
THE ADVENTURE OF THE EMPTY HOUSE

THE END OF THE 1930S WAS A TIME OF TRANSITION AND TRAVEL.

Albert Szent-Györgyi was immersed in transition and travel—all of it distinctly unwanted. He spent the early part of the decade on the faculty of Szeged University in Hungary, where he probed the fascinating mystery of muscle metabolism. But as the 1940s approached, the scientist became dangerously outspoken in his anti-Nazi, antifascist politics. Szent-Györgyi vigorously opposed Hungary's increasingly sinuous alliance with the Axis powers, and he took public, often unpopular, stands on behalf of his beliefs. Anti-Semitism was growing in eastern Europe, where even liberal academic campuses like Szeged were not immune to the metastasizing politics of bigotry. Risking his own safety, Szent-Györgyi used whatever influence and connections he had to help scientists like Hans Krebs (future 1953 Nobel Prize winner in Physiology or Medicine) and other Jewish colleagues avoid persecution or, if necessary, escape the country.[1]

In 1943 Miklós Kállay, prime minister of wartime Hungary, sent Szent-Györgyi to Turkey; officially, he was supposed to deliver a series of scientific lectures in Istanbul. In reality, Albert "Spy"-György was initiating secret discussions with representatives of the Allied governments. When the Gestapo uncovered the scientist's involvement with the enemy, Adolf Hitler personally issued a warrant for his arrest. Upon returning to Hungary, Szent-Györgyi was placed under house arrest, but he managed to escape to the Swedish embassy in Budapest.[2] He was given refuge there and eventually received Swedish citizenship. When the opportunity arose, Szent-Györgyi was folded into the tiny trunk of a rust-eaten eastern European automobile and smuggled out of the embassy.[3] He moved for months between various safe houses in several Hungarian cities, hiding until Russian troops eventually liberated the country. Once his exploits during the war became known to his compatriots, Szent-Györgyi was transformed into a national hero.[4] There was even a short-lived movement to have him run for president of Hungary.[5]

Szent-Györgyi's wartime sacrifices extended beyond the physical inconveniences he suffered, or the dangers he faced on behalf of his friends, colleagues, and country. He paid for his antiwar stance in other ways, some of which involved serious financial repercussions. Lots of people are all talk. But Szent-Györgyi backed up his soapbox spiels with his pocketbook:

> "When I received the Nobel Prize," wrote Szent-Györgyi, "the only big lump sum of money I have ever seen, I had to do something with it. The easiest way to drop this hot potato was to invest it, to buy shares. I knew that World War II was coming and I was afraid that if I had shares which rise in case of war, I would wish for war. So I asked my agent to buy shares which go down in the event of war. . . . I lost my money and saved my soul."[6]

John F. Kennedy was also dealing with transition and travel. Between 1938 and 1940, while he was a student at Harvard, the future president's problematic digestive tract still wasn't behaving itself.[7] He returned to the Mayo Clinic, but no cure was forthcoming. A subsequent two-week stay in New England Baptist Hospital was likewise not helpful. In February 1939, despite his intense and growing hatred of hospitals, Kennedy agreed yet again to return to the Mayo Clinic for follow-up evaluations. It turned out to involve more of the same: rest, bad food, and a thick proctoscope. One can only marvel at the charm and confidence Dr. Alvarez must have exuded to persuade the young Bostonian to keep returning to Rochester—especially when JFK seemingly received so little that was helpful in return.[8]

If Kennedy was indeed taking DOCA at this time, the drug didn't seem to be soothing his colitis appreciably—worse, it was probably contributing to new problems. For example, it is classically taught (although still somewhat controversial) that steroids like DOCA can acutely produce or aggravate stomach and duodenal ulcers. Although JFK was not diagnosed as having gastroduodenal ulcers until the fall of 1943, it's probable that his ulcers were beginning to arise at this time;[9] if so, they may have contributed to his pain and weight loss. Incredibly, Kennedy appears to have been the first prominent person to suffer significant side effects from steroid use.

Unbeknownst to Kennedy, or to the handful of doctors prescribing these primitive steroids, DOCA had the potential to do something else. Something that would eventually affect almost every aspect of JFK's life. Something really bad.

It turned out that DOCA could *cause* Addison's disease.

Nineteen thirty-nine was, in a perverse sense, also a time of transition for the Mayo Clinic. William and Charles Mayo, who as young men had raced from the great Rochester tornado of 1883, both died.

Dr. Charlie was the first to go, passing away at the age of seventy-three on May 26 in Chicago—the city in which he had attended medical school. The cause was pneumonia. A physician familiar with the case was quoted in the *Rochester Post-Bulletin*; he described Dr. Charlie's condition as a rare form of "type three plus type six pneumonias."[10] "Gentle, kindly, Dr. Charlie," as he was described in his eulogy—a man whose looks, quips, and behavior reminded others of Will Rogers—had been the heart and soul of the clinic's leadership. Only he could have suggested that "there is no fun like work" and truly meant it. Dr. Charlie's death was a blow to the spirit of the Mayo Clinic.

On July 28, 1939, two months after the passing of his brother, Dr. William Mayo died.[11] He was seventy-eight. He had suffered from abdominal discomfort for months; it had been intermittent at first, but eventually became continuous and unrelenting. After returning urgently to Rochester from his winter home in Arizona, William Mayo underwent surgery at Saint Marys Hospital. The findings were devastating; he had widespread cancer. When he died shortly after the operation, the actual loss to the clinic was more emotional than physical: like his brother Charles, William Mayo had not been involved in the day-to-day operation of the clinic for many years. But he had been the "brain" of the institution and, when necessary, the "hammer" that kept things in order. His guidance and strength, even if only symbolic, would be missed.

Life went on, but the pall cast over the Mayo Clinic by the nearly simultaneous deaths of its founders lingered on for years.

Nineteen thirty-nine likewise meant transition and travel for Philip Hench. But there was, of course, nothing unusual about that—Hench always seemed to be transitioning and traveling. He had just accepted the presidency of the American Rheumatological Association and would soon become president of the American Rheumatism Association, an organization he had helped to found. Hench kept busy with his clinical practice and research projects; his work continued to focus on patients with jaundice, pregnancy, or other forms of stress-induced remission of rheumatoid disease.

But Hench was also increasingly interested in the history of yellow fever.[12] While his friend Ralph Hutchison seemed more determined each day to hold a Founder's Day celebration in honor of the Washington and Jefferson College's most famous alumnus, Jesse Lazear, Hench was increasingly determined to elucidate the yellow fever story in which Lazear had played such a big role. The Mayo rheumatologist spent increasing amounts of time collecting documents about Walter Reed and his coworkers, corresponding with survivors of the project, and, as improbable as it seems, watching movies about yellow fever.[13]

By the fall of 1939 Hench was spending considerable time and effort helping "Hutch" plan for the Washington and Jefferson Founder's Day festivities. Hench wanted his friend to line up the best possible keynote speakers for the event. He suggested inviting some recent Nobel Prize winners, although he conceded that "most of them are not very good speakers."[14] Specifically, he acknowledged that William Murphy (the Harvard laureate now caring for JFK) was a "very poor speaker." The same went for Murphy's partner and co-Nobel laureate, George Minot, whom he called "only a fair speaker—mutters and mumbles" (quite a criticism from a man with his own significant speech impediment). He considered inviting Banting, the Nobel winner for insulin, but pointed out that he was currently in the army, and "you don't want to count on him because he might be in Europe at that time." (Banting, as noted earlier, died shortly after this assessment when his plane crashed on its way to Europe.) He thought that Charles Best, Banting's partner in the insulin discovery, might be a good invitee even though he had not won the Nobel Prize; although only thirty-five years old, the young scientist was now very well known—and a good speaker. But Hench thought that Best was "not in the class with the others." He considered a German scientist, Gerhard Domagk, to be an excellent choice for keynote speaker even though he, like Best, wasn't a Nobel Prize winner. Domagk's work with prontosil, the sulfa drug that had emerged as the first antibiotic, was now extremely "hot" in medical and scientific circles.[15]

Finally, Kendall was engaged in transition and travel.[16] In 1938 he attended the International Congress of Physiologists in Zurich and had the opportunity to meet his main competitor in the race for cortin, Tadeus Reichstein. He described the meeting as "a pleasure." The two discussed DOCA (the compound that Reichstein had recently synthesized for the first time), including some of its physiological aspects with which they, as chemists, were not terribly well versed. Although neither man was a physician, they both recognized that DOCA helped "a little" in Addison's disease, and that it had mostly an effect on salt and water metabolism. But it did not prevent death or make patients feel better. There was mutual agreement between the two rivals that it was not *the* cortin.

Kendall wouldn't be making another business trip to Europe for twelve years.

CHAPTER 12

War Looms

Though unmusical, German is the most expressive of all languages.

—SHERLOCK HOLMES, "HIS LAST BOW"

THE WORLD WAS LEARNING JUST HOW "UNMUSICAL" GERMAN COULD be. World War II began with the invasion of Poland on September 1, 1939. British prime minister Neville Chamberlain, in an address later that day, summed up the sense of hopelessness much of the world was feeling at the moment:

> We have no quarrel with the German people, except that they allow themselves to be governed by a Nazi government. As long as that government exists and pursues the methods it has so persistently followed during the last two years, there will be no peace in Europe. We shall merely pass from one crisis to another, and see one country after another attacked by methods which have now become familiar to us in their sickening technique.[1]

Although the war would be fought, for now, in Europe, the Americans were not indifferent to the threat posed by Germany's "sickening technique." Before the onset of the first blitzkrieg, the United States was keenly interested in all things related to the Nazi military machine. In early 1940 a rumor began to surface that caused particular concern.

Since the days of Addison and Brown-Séquard, scientists believed that "cortin" from the adrenal cortex was essential for helping the body resist stress. In its absence (that is, after adrenal gland removal), animals became fragile and delicate; even minimal stressors like a sudden loud noise could kill them. It stood to reason that extra adrenal cortical hormone might help animals, including humans, survive stress—perhaps even stresses that would normally incapacitate or kill. It was in this setting that a certain rumor began to circulate: the claim was that the Germans had isolated cortin, and that the Luftwaffe was using it to help pilots fly at higher altitudes.[2]

It was widely quoted that the adrenal extract counteracted the effects of hypoxia and allowed flights to 40,000 feet or more.[3] If true, this was an aviation miracle—and a significant threat to any armed forces that might oppose the Germans. These rumors spawned other rumors, the most concerning of which was that German submarines were secretly traveling back and forth between Argentina and Germany, transporting large amounts of beef adrenal glands for the sole purpose of preparing cortical extract. Although these rumors were never proven and, in retrospect, were almost certainly untrue, their mere existence demanded immediate action by the Americans.

In 1941 the National Research Council of the United States met and set national priorities for government-sponsored research. The impending, seemingly inevitable conflict with Germany had no small role in defining these priorities. Third on the list was the development of new antimalarials, which would be essential for supporting combat activities in tropical zones. The second priority was the creation of a program to produce penicillin, which was already seen as a miracle drug and was obviously critical to a war effort. The number one priority was the identification, isolation, and synthesis of the hormone of the adrenal cortex—cortin.[4]

Incredibly, as America teetered on the brink of entering World War II, it seems the U.S. government believed it necessary to meet the Nazis head-on in the race to develop cortin. A committee of fourteen internationally recognized chemists was assembled to direct this research: William Mansfield Clark of Johns Hopkins chaired the committee, which included James Collip of insulin fame, as well as the head of the Mayo Clinic's biochemistry lab, Edward C. Kendall. "Nick" Kendall was no closer to solving the mystery of cortin, but the search had just gone from the back alleys of science to front and center stage.

Walter Alvarez, now fifty-seven years old (two years older than Kendall), could offer little more than moral support in the impending war. Apparently believing that "the pen is mightier than the sword," he continued his prolific writing and publishing. Many of his 347 articles, 329 editorials, and books/ monographs (one of which, *Nervous Indigestion,* sold over 35,000 copies) were written or rewritten during this time. In 1938 he became the editor of the *American Journal of Digestive Diseases,* and in 1941 he started a new journal called *Gastroenterology*; it remains the top-rated journal on digestive diseases today. Walter's sister, Mabel Alvarez, also served her country during the war by volunteering for the Red Cross.[5] Mabel eventually worked at the Naval Hospital in Long Beach, California, where she employed her artistic skills rehabilitating injured sailors.

But Walter's wandering son Luis, now a research physicist in Southern California, was perhaps the least likely Alvarez to make a significant contribution to the war.

Following his move to Berkeley, Luis Alvarez stopped behaving like a Rochester tornado and began thinking like one. His California colleagues quickly took to calling him "prize wild idea man" because of his interest and abilities in so many different aspects of physics.[6] Although his first love had been light and optics, Alvarez quickly migrated to the growing field of nuclear physics. His work included the demonstration of electron capture by nuclei, the creation of an early neutron beam, studies on the magnetic properties of neutrons, the development of a mercury lamp, and seminal work on the tritium atom. Despite the fact that this was real science performed on real particles and waves, the immediate value of this research was largely theoretical. There was no practical application for most of these things, and certainly no military application. It therefore seemed overly optimistic—or perhaps pointless—when the military tapped Luis in 1940 and sent him to the Massachusetts Institute of Technology (MIT) Radiation Laboratory.[7] There was little reason to predict that the "prize wild idea man"—the rebellious teen from Rochester, the son of righteous Dr. Alvarez, the nephew of America's quirkiest female artist—would contribute anything meaningful to a country at war. But this view of things would turn out to be wrong.

Within his first few days on the job at MIT, Luis was doing research on radar. His accomplishments in the next three years were impressive and may have changed the course of the war. Luis developed three entirely new radar systems. The first, based on his weather-plagued flight into Boston, used a narrow radar beam to enable a controller on the ground to blindly guide an airplane into a landing—this is still the way airplanes land when the pilot's direct vision is inadequate. The second radar system, which went by code names like "Eagle" and "Vixen," enabled pilots to locate and identify objects on the ground (or submarines in the water) that were hidden by clouds or darkness. Bombing by radar became a reality. The third system, which utilized microwave radar, provided an "early-warning" capability for identifying aircraft that were obscured by dark or overcast conditions.[8]

Luis Alvarez's breakthroughs in radar technology were stunning accomplishments and gave the Allies a huge advantage over their military adversaries. Luis could have returned to Berkeley in 1944 with the satisfaction of knowing he had made a critical difference in the war. But he didn't. He had one more military contribution to make.

Ernest Hemingway completed *For Whom the Bell Tolls*, a novel about the Spanish Civil War, in 1940 after three years of research and writing. The book was an instant hit. While still in Madrid he met Martha Gellhorn, a war correspondent; she became his third wife. Before they were married, the couple had been traveling back and forth to Cuba, where they usually stayed at rundown places like the Hotel Ambos Mundos. Martha was tired of shabby abodes. She dragged Ernest back to Cuba, and from a newspaper advertisement identified a fifteen-acre farm outside of downtown Havana called Finca Vigía (Lookout Farm). They initially rented it for $100 a month, and even at that relatively low price it was no steal. Rundown and decrepit, Hemingway thought it was not worth trying to renovate. His new wife felt differently. Acting as her own contractor, she hired workers and began remodeling.[9]

Martha Hemingway must have done something right in her efforts to fix up Finca Vigía. Ernest, who had never owned a home and had never wanted to settle anywhere, apparently liked what she'd done. With its restored swimming pool and tennis court, Hemingway, along with a fleet of cats and dogs that roamed the place, finally felt at home. Best of all, he was within quick striking distance of Havana's Floradita Bar, where he had a reserved seat at the end of the bar from which he could slug down frosty daiquiris and mojitos. Armed with money from his recent literary successes, Hemingway "bought the farm" he'd been renting. The selling price was $18,500.[10]

Once settled, Hemingway, now an international celebrity, began working on his next book, *Across the River and into the Trees*. It would not be published until 1950. Most critics consider it a weak piece of literature. But Hemingway was also simultaneously working on another novel, which wouldn't be published until 1952. This book was going to leave a much different impression on the world.

As Martha Hemingway was knocking down Cuban walls, Philip Hench was preparing to go to Havana and search for Camp Lazear, the site of Walter Reed's experiments on yellow fever.[11] His trip was delayed for several weeks when his son Kahler needed an emergency appendectomy.[12] Once the crisis had passed, Hench left on what he expected to be a pleasant hobby-related vacation. It didn't exactly work out as planned.

Hench's trip to Cuba, conducted with the blessing of the Cuban government, started uneventfully. As he put it, "I went down there expecting to investigate the story for two days, and to loaf for eight days."[13] But his itinerary changed when he was taken to the place that the Cuban government had identified as Camp Lazear. The government was planning to erect a memorial building on the site. Hench, largely because of his meticulous research done in preparation for the trip, quickly realized the site was not correct. The Cuban government had misidentified the location of Camp Lazear.

In the span of ten days, Hench accomplished what he often accomplished—the impossible. Relying on his exquisite knowledge of the Walter Reed story, he began to ferret out the real location of Camp Lazear. According to Hench,

> I got involved in quite a controversy. I soon found out that the campsite that has been marked by the government and where they plan to have a memorial building, which has been named Camp Lazear, is not the proper campsite at all. After considerable work I located it a mile away. Then, I had to try to prove it, and I had several conferences with the secretary of war and the secretary of health, with their respective staffs, as well as the officials of the Finlay Institute. I had to get documentary proof and also collected a certain amount of photographs taken by myself, others in the form of professional photographs, . . . suffice it to say I think I finally convinced the government officials that I was right and they have asked me to write the matter up, forward it to the Cuban government, and they will appoint a commission to determine the final site.

Hench, understating as usual, commented, "I never worked so hard in my life, but I had a very interesting time."[14]

Ernest Hemingway had gone to Havana to settle down, relax, drink, and write. Hench had gone to Havana on a pilgrimage to the site of his idol's great triumph. And for ten brief days in the spring of 1940, these two giants, destined to be linked to one another by the Mayo Clinic and the Nobel Prize, were a few miles apart in Havana chasing their respective dreams.

CHAPTER 13

Hench Meets Kendall

Local aid is always either worthless or else biased.

—SHERLOCK HOLMES, "THE BOSCOMBE VALLEY MYSTERY"

THE JAPANESE ATTACKED PEARL HARBOR ON DECEMBER 7, 1941. The United States was now knee deep in a war on two fronts.

Luis Alvarez was doing his part at MIT, developing the radar systems that would prove critical to the war's outcome. But as the spring of 1941 approached, all he could develop was abdominal pain.[1] Mild and intermittent at first, the pain slowly worsened. Meals, especially fatty ones, became torture. By the time Luis sought medical help, the situation was becoming urgent. Tests confirmed that his gallbladder was the problem. It had to be removed. Luis was living in Cambridge, Massachusetts, a community with excellent medical care, but his father was one of the world's preeminent gastrointestinal (GI) doctors, and his opinion mattered—and according to Walter, if Luis was going under the knife, it was going to happen at the Mayo Clinic. Luis returned to Rochester immediately.

The operation, a cholecystectomy (gallbladder removal), was performed by Dr. Waltman Walters, one of Mayo's best-known surgeons and, perhaps not coincidentally, the son-in-law of Dr. William Mayo.[2] (As a surgical resident in training, Dr. Walters married William Mayo's youngest daughter, Phoebe.) By any measure, Walters went on to become a superb physician and surgeon. But some observers believed his position in the clinic's pecking order was higher than it should have been based on his administrative talents.[3]

Unfortunately, Luis's surgery did not go well.[4] Complications developed. He spent seven weeks in bed, and during this time he developed a blood clot, or deep vein thrombosis, in his leg. This occurrence sometimes leads to pulmonary embolism, a potentially deadly situation in which the blood clot breaks loose from the leg and travels to the lung, where it lodges and blocks blood flow. Luckily, by 1941 there was a radical new treatment for leg clots.

Dr. Charles Best, the young medical student who had assisted Banting in the discovery of insulin, was (as Hench had implied when he suggested the Canadian as a potential keynote speaker for the Washington and Jefferson College's Founder's Day celebration) now a talented researcher in his own right. His most recent discovery was a new drug for treating blood clots; it had been first derived from the liver (or "*hepar*" in Latin) and was therefore called heparin. This substance, which could be obtained from animal livers or lungs, stopped blood from clotting and prevented the lethal complications of venous thrombosis. Luis almost certainly received this drug at Mayo. Just as important, the previously mentioned Dr. Hugh Butt—once a medical resident shivering in Dr. Ancil Keyes's walk-in refrigerator—was now a full-fledged Mayo staff member who was experimenting on patients with a substance made from spoiled sweet clover that could be taken by mouth and would likewise prevent clotting. This drug, originally discovered at the University of Wisconsin and called Coumadin or warfarin (for Wisconsin Alumni Research Foundation), was just starting to make its way into clinical medicine. Dr. Butt would report his successes in a 1941 publication of the *Proceedings of the Staff Meetings of the Mayo Clinic.*[5] Luis Alvarez might have received Coumadin as part of his treatment—perhaps his case was one of those reported in Butt's landmark communication.

These cutting-edge medical advancements for treating blood clots, which fortunately were made just as Luis Alvarez needed them, may well have saved his life. He slowly recovered from his operation and eventually returned to MIT, where he completed his work on radar.

Unknown to Luis, the government was assembling a group of top-notch physicists who would soon begin working on something secret called the "Manhattan Project." The young scientist from Rochester had caught their interest.

With regard to the war against Germany and Japan, Philip Hench could see the writing on the wall. He, along with a number of other Mayo Clinic specialists, volunteered for the army. Specifically, "eighteen of us are offering our services to the Surgeon General as specialists to be used if necessary."[6] Hench knew his offer would be accepted, so it was just a matter of time until he went to war.

And before he left, there were a few things he needed to wrap up.

One loose end involved his rheumatology research. As noted previously, for the past ten years Hench had been making detailed observations on patients with active rheumatoid arthritis, especially those who went into spontaneous remission. He'd first observed this phenomenon occurring in a patient during a bout of jaundice, but he had subsequently witnessed it with pregnancy, starvation, injection of typhoid vaccine, and general anesthesia (even when surgery wasn't performed).[7] Convinced that the liver must be playing a role in this, Hench was testing ways to mimic the therapeutic effects of jaundice. Considerations included intravenous infusion or oral administration of things like cholesterol, various bile acids from humans and cattle, assorted liver extracts, and even blood from jaundiced patients. In several cases Hench gave these agents to his patients and successfully turned them yellow, but so far he had been unable to relieve any rheumatological symptoms by doing this. He needed to do something more. What? Hench already thought he'd gotten in a little over his head by administering some of these agents; many were rare, potentially dangerous substances. There might be other things he could try, but he was not familiar with their properties. And, if nothing else, Hench was concerned about the patient's safety. As his associates pointed out, "we always gave appropriate attention to basic humanistic principles . . . all of which were applied voluntarily in our studies."[8] Hench would not be comfortable subjecting his patients to any more jaundice-producing or body-stressing agents until he knew more about them. He needed to talk things over with someone. Who?

That "someone" was Edward Kendall. Hench had, of course, met Kendall before—they'd both been on the staff of the Mayo Clinic for more than a decade. But they had not interacted scientifically. Hearing of the senior chemist's current interest in steroid research (it was known at the time that bile acids and other jaundice-producing substances are chemically related to steroids), Hench sought out Kendall and asked his advice about the various agents. The discussions that ensued were earnest and fruitful. Kendall and Hench went on to meet many times during 1941, and a number of experimental options for treating rheumatoid arthritis were discussed. One of them involved "compound E."

As noted earlier, Kendall had already isolated a number of different compounds from the adrenal cortex—these were identified as compounds A through F. Dwight Ingle, using his rat muscle assay, determined which of these were biologically "active." Compound E demonstrated the most profound effect on the muscle assay; even if it turned out that "E" wasn't *the* cortin, it was nonetheless a compound of great interest. The problem?

Compound E wasn't available in quantities large enough to allow experimentation, especially on humans.

Kendall and Hench both thought it unlikely that a "mineral corticoid"—that is, a steroid substance like DOCA that primarily affected the retention or excretion of salt and water—would be effective in the treatment of rheumatoid arthritis. But what about something like compound E that influenced metabolism? Perhaps this substance would have an effect on inflamed joints as well. Hench knew nothing about compound E. Kendall knew nothing about rheumatoid arthritis. But between the two of them, a "fragile but logical" decision was apparently reached to try compound E in patients with rheumatoid arthritis when it was finally available in sufficient quantities.[9]

Hench's junior associate, Howard Polley, subtly questioned this account of the decision and implies that Hench may have embellished the story years later. According to Polley, "[Hench] stated on many occasions that he had made [a note in his notebook] regarding his January, 1941 conference [with Kendall]. *Decision made to try "E" in RA.* . . . We consider it curious that neither we nor any of our colleagues nor even P. K. Hench [Hench's son] had ever seen this notation."[10]

Conan Doyle, like many new authors, found it hard to get his first book published. After three publishers had rejected *A Study in Scarlet,* he finally got the London firm of Ward Lock and Company to pay £25 for the right to print it in the 1887 edition of *Beeton's Christmas Annual.* It was a paltry sum, but it launched the author's career. He soon became a prolific writer.

Now it was Hench who was writing like Doyle on amphetamines. In addition to the multiple letters he composed every day, either due to medicine-related issues or because of his hobbies, he was also trying to finish the "Seventh Annual Rheumatism review which I edit yearly and write (90 percent of it) myself. It is due in three weeks and is 300 pages long!!"[11] Between his writing, research, clinical duties, and impending military duty, it was indeed a hectic time for Hench.

Yet through the stress Philip Hench's sense of humor was always evident. He had been awarded an honorary doctorate from Washington and Jefferson College during the Founder's Day honoring Lazear. The degree had been presented by his best friend and president of the college, Ralph Hutchison. In the aftermath of the ceremony, a frisky Philip Hench took a little pleasure in tormenting his close friend. Kidding him about the failure of the college photographer to capture the moment when the diploma was conferred, Hench wrote, "Being a photographic fiend, I am really sorry that it so happened that I was the only one not having a photograph taken when you presented the degree. Did you by any chance tell the photographer, 'for heaven's sake, don't photograph Hench. We don't want any permanent record of that!'?"[12] His friend had also made the entirely forgivable mistake of introducing Hench as the president of the "World Arthritis Association" rather than the "American Rheumatism Association." Hench wrote to Hutchison afterward, "also did you remember my request that if for any reason my citation should be published anywhere in your records, please demote me from president of the all-universe, all-world, all-hemisphere Rheumatism Association to merely the president of the American Rheumatism Association."[13]

Maybe Hench took a little too much pleasure in chiding his chum? Surely he knew that "what goes around, comes around." Ralph Hutchison was planning his revenge. Shortly after the ceremony, Hench made a contribution of $150 to the Washington and Jefferson College fund. Being a graduate of Lafayette College, Hench sent the money to Hutchison with a note jokingly stating that he wanted it clearly understood this check was coming from a "Lafayette Class of '16 graduate" and "not a Washington and Jefferson alumnus" (having just gotten his honorary degree, he was now technically an "alumnus"). A month after the ceremony, when this insignificant jotting had been apparently forgotten by all, "Hutch" sent his friend a letter: "Dear Phil, Christmas news! You've gone to town! . . . I sent that letter on . . . with your pledge and a check . . . and darned if (they) didn't take you literally."[14] Hutchison went on to explain that a souvenir book commemorating the Founder's Day ceremonies was on its way to 5,000 distinguished parties across America. Hutchison had enclosed a copy of the book for Hench. One of the pages was dog-eared; flipping to it, Hench instantly saw his quote spread across the top of the page. In the cold absence of context, it looked harsh and ungrateful toward the college that had just honored him so graciously. "Please note the page marked!" wrote Hutchison. "Well, you are famous anyway!"[15]

A panicked Hench telegraphed his friend immediately: "Afraid my humorous personal remark boomeranged badly. Greatly concerned less comment printed so boldly will initiate among alumni and college friends comments unfavorable to us both. . . . nothing flatter than misplaced humor. Subtle compliment was intended but cold print looks bad. Beg you stop distribution."[16]

Hench received a telegraph reply the next day. "Thanks for your wire which will be framed. Will now advise that this was a special edition of one copy devised and printed . . . as *a Hench joke on Hench.* Apologies and Merry Christmas."[17]

Hench, humbled by the practical joke played on him, wrote back: "Boy did I swallow that one. You're ten up on me. To think I've got a science degree and couldn't smell that one. Next time I come I'll bring along wet nurse to feed me pure milk. . . . I believe in Santa too."[18]

Thirty-year-old Russell Marker wasn't going off to any foreign war. But that didn't mean he wasn't about to do a lot of traveling.

Marker, having discerned how to make progesterone out of costly sarsasapogenin, was desperately trying to find a less-expensive starting material. After studying a little botany, he realized that certain gourds, squashes, and yams contained a variety of primitive steroid precursors similar to sarsasapogenin; many of these were capable of undergoing the desired conversion. Specifically, he zeroed in on a sapogenin called diosgenin. According to Russell's calculations, this well-known, but rarely encountered, naturally occurring substance could be converted with the fewest steps into progesterone. He therefore needed to find a plant that contained a high concentration of diosgenin. Searching woodlands from the north of Canada to the south of Mexico, he finally struck gold near the city of Orizaba in the state of Veracruz, Mexico. An analysis of the wild yam *Dioscorea*, a gnarly black tuber unique to the region, proved it to be a potent source of diosgenin.[19]

Marker eventually collected two large sacks of the tubers in the mountains around the city of Veracruz. He put them on the top of the bus in which he was traveling and returned to Orizaba. When he arrived, the bags had disappeared! It was only by bribing a local policeman that he was able to recover a single 50-pound root.[20] Smuggling the illegally obtained agricultural product across the border, Marker brought it back to his laboratory in Pennsylvania and quickly demonstrated that diosgenin could be extracted from the yam.

CHAPTER 14

World War II and Military Steroid Research

There's an East wind coming all the same, such a wind as never blew on England yet. It will be cold and bitter, Watson, and a good many of us may wither before its blast. But it's God's own wind nonetheless and a cleaner, better stronger land will lie in the sunshine when the storm has cleared.

—SHERLOCK HOLMES, "HIS LAST BOW"

WORLD WAR II WAS STIMULATING RESEARCH IN STEROIDS. Unfortunately, it was often the ugliest research imaginable, and as the 1930s came to a close the cortin story was taking a short, unpleasant digression down the wrong path.

Rumors continued to buzz about Nazi interest in cortin and the possibility that the Nazis were isolating the substance from Argentinean beef adrenal glands and using it to help Luftwaffe pilots tolerate the stress of high-altitude flight. Similar rumors soon began circulating about Nazi interest in anabolic steroids.

Substances like androsterone and testosterone, both steroids, produce so-called anabolic effects—development of muscles, emergence of body hair, deep voice, and other masculine qualities. These steroids are also known to increase aggressiveness. As the war progressed, it was alleged that these "Nazi drugs" were being used to "pump up German soldiers."[1] There is no evidence that this ever occurred.[2] Not that it mattered; as with cortin, the simple existence of these claims was enough to generate international concerns in government and scientific circles.

Although the rumors about pilots taking cortin and soldiers taking anabolic steroids were likely not true, the Nazis were definitely interested in—and experimenting with—various steroids. Were they working on cortin? Maybe. Estrogen and certain female hormones? Probably. Androgens and other male anabolic steroids? Certainly. Unfortunately, much of this "research" was taking place in concentration camps, with unwilling prisoners as the experimental subjects.

As noted earlier, in 1930 Dr. Carl Clauberg developed an assay for progesterone (a key precursor in the manufacture of cortisone) that allowed, for the first time, the potency of various progesterone-containing compounds to be measured. Progesterone, a steroid manufactured in the ovary, plays a crucial role in pregnancy and the development of the fetus (in contrast to the ovary's other major steroid hormone, estrogen, which is primarily involved in the development of female sex characteristics). Progesterone and estrogen interact in complex ways to facilitate reproduction and general "femaleness." Clauberg, a professor of gynecology at Koenigsberg University in Germany when the war began, was among the world's top experts in the study of these hormones. A fanatical advocate of the Nazi Party, Clauberg opted to use his knowledge and ability in support of the party's ideology. In 1942 he asked Heinrich Himmler if it would be possible for him to study various ways to sterilize the "undesirables" that Germany was attempting to eliminate from Europe. Himmler, already familiar with Clauberg's expertise in reproductive hormones, accepted the offer. In December 1942 he sent Clauberg to Auschwitz where, under the guidance of the infamous Angel of Death—Josef Mengele—the misogynist gynecologist established a research facility in Block Number Ten of the main camp. His mission: to find methods for "cheap and efficient" sterilization of women.[3]

The subsequent "experiments" carried out under Clauberg's guidance had almost nothing to do with science; they pandered to perversion and camouflaged cruelty. The data from these pseudostudies are not worth examining. But based on firsthand reports from survivors, there can be little doubt that Clauberg's initial experiments involved the use of primitive steroid hormones.

> Test fluid sterilization experiments were performed by Professor Clauberg or under his control. He was an eminent German gynecologist and acted under Himmler's orders. According to one of the witnesses Professor Clauberg admitted that his experiments were of no scientific value. Identical results were previously obtained on animals and were well known to the medical profession.

Thus the experiments on women in the Auschwitz camp could not serve any scientific end. In addition they were performed in terrible conditions which often led to chronic illness, permanent injury or even death. Neither the doctors nor the assistant personnel were properly trained for the purpose. Unsterilized instruments and dressings were often used. . . . Injections of hormones to women were also made and results observed.[4]

It can be inferred that hormonal manipulation by injection of unknown agents (presumably related to sex steroids) did not produce the desired effects, because the experiments soon progressed to involve more grotesque, nonhormonal techniques. "Sterilization of women was carried out by the pumping of a thick white test fluid, consisting of contrast medium and some unknown chemical agents, into the uterus and tubes. Also sterilizing operations were performed, the uterus, tubes and even sometimes breasts being removed. Women experienced great suffering during test fluid experiments."[5]

When these approaches also proved unsatisfactory—perhaps because they were too slow, too difficult, or too unreliable—Clauberg switched to other methods for destroying female reproductive capability. These included bombarding his subjects with x-rays. It is unknown how many prisoners were killed by these methods, but it is estimated that 700 somehow survived the crude sterilization procedures. Many of these survivors were later euthanized for the purpose of autopsy.[6]

Within two years Clauberg believed he had developed an acceptable technique for sterilization of women; to what extent, if any, this approach involved the use of steroids remains unknown. He may have given up on steroids like estrogen and progesterone, concentrating instead on physically destroying the reproductive organs with radiation or direct mechanical injury.

In June 1943 Clauberg sent Himmler a message suggesting that the experiments were proving successful:

The non-surgical method of sterilizing women that I have invented is now almost perfected . . . if the research that I am carrying out continues to yield the sort of results that it has produced so far (and there is no reason to suppose that this shall not be the case), then I shall be able to report in the foreseeable future that one experienced physician, with an appropriately equipped office and the aid of ten auxiliary personnel, will be able to carry out in the course of a single day the sterilization of hundreds, or even a thousand women.[7]

Clauberg's twisted interests in steroids resulted in unspeakable evil directed at women.[8] Another Nazi doctor, Carl Peter Vaernet, also harbored an unhealthy interest in sex hormones. But these hormones were different from those studied by Clauberg, and the victims of his equally cruel research were likewise distinctly different.

Vaernet was born in Denmark near the city of Aarhus in 1893; he was the son of a wealthy trader.[9] He began experimenting in endocrinology in 1932. His field of interest? Vaernet was obsessed with and/or terrified by gay men, and orchestrated a public campaign to "fix" them. He believed that hormonal manipulation, using the newly discovered steroid testosterone, could "cure" these "unfortunates."

Vaernet began to study this subject seriously in 1939, and by 1941 newspapers were starting to cover his admittedly fascinating animal work involving sex hormone manipulation. There were tales of hens running around outside of his clinic crowing as if they were roosters—a behavioral transformation attributed to hormonal interventions.[10] At first Vaernet's experimental work was balanced against his clinical practice; just as Mabel Alvarez's beau, Dr. Robert Kennicott, had been one of Beverly Hills premier "society" doctors, Vaernet was likewise a "high society" doctor in Copenhagen. But as war ensued and it became widely known that he was a close friend of Frits Clausen (another Danish doctor who headed the Danish Nazi Party), Vaernet's popularity with his peers—and his medical practice—soured.[11] Dr. Vaernet quickly found himself with a lot fewer patients—and a lot more time on his hands—to pursue his increasingly bizarre interests in the treatment of homosexuality.[12]

Vaernet believed that homosexuality resulted from a deficit in testosterone and that correcting the deficit would eliminate the problem. In 1942 Vaernet began experimenting with homemade testosterone implants.[13] One patient, a gay schoolteacher, was described as having a "good result" from his implant; he eventually married.

Realizing that the available methods for testosterone administration were far from perfect, Vaernet tried to design a better apparatus for hormone delivery.[14] In 1943 he patented a device he'd designed specifically for this purpose.[15] The contraption was an open-ended metal tube packed with testosterone. Once implanted, the steroid leaked out of the tube over several years, making sustained-release of testosterone possible. Implantation was performed on the right side of the groin because, according to Vaernet's convoluted theories, this was supposed to produce a greater change in the patient's sex drive.

In 1944 Vaernet raised the stakes by joining the SS. He obtained a position at the Buchenwald concentration camp, where he worked exclusively on experimental cures for homosexuality. This move had the approval of not only Himmler and Reich medical chief general Dr. Ernst Grawitz but also the endorsement of his immediate director.[16] Vaernet was given the rank of major by the SS along with ample funding, laboratory facilities, and a free run of the prisoners to use as experimental subjects. The Danish Nazi showed exceptional enthusiasm for his work.

Prison camps have a way of changing those who control them. If there were ever any vestiges of decency in Vaernet's life, they disappeared at Buchenwald. His behavior became savage and his experiments gruesome.[17] The inmates included a large number of purportedly gay men; these individuals were marked with pink triangles to identify them and indicate their suitability for medical experimentation. Vaernet started his work on familiar ground—by implanting his testosterone-filled metal tubes in order to raise their hormone levels. Other SS doctors at the prison joked about this operation and called these implants "firestones" or "flintstones."[18] They remained skeptical that steroid implants could change a subject's sexual persuasion. When these devices failed to produce the desired results, Vaernet upped the stakes. He began experimenting with castration as a way of curing homosexuality. Between thirteen and fifteen gay men were operated on, and it appears most of them died from the surgery.[19] Ultimately, Vaernet's research resulted in . . . nothing. "No positive findings were ever obtained."[20]

The cortisone story is the high point of a bigger story encompassing the history of all steroids; it's hard to imagine a lower moment in any chapter of the steroid saga than the sordid tales of Clauberg and Vaernet.[21]

The importance of "pilot stress" and its impact on combat aviation cannot be overemphasized; the Allies and Axis forces were both desperate to solve flight-related problems and improve pilot performance. If cortin wasn't the immediate answer to high-altitude, high-stress aviation, what was? Once again, the Nazis were not adverse to the use of imprisoned human subjects for research purposes.

In the Dachau concentration camp during the middle of 1942, a hypobaric (low-pressure) chamber was built and tested.[22] Experimental subjects were placed inside the chamber, and the air was pumped out of it; the partial vacuum created by this process simulated the low atmospheric pressure encountered at high altitudes. During these studies subjects typically lost consciousness—and occasionally died—from the stress.[23] The Nazis used this apparatus to predict human tolerance and performance at high altitudes. They studied various approaches to modifying these responses (and if cortin had been available, its ability to increase high-altitude tolerance could have been assessed using the chamber). But it was bad science, and bad science is its own reward. As was the case with the sterilization experiments, it is doubtful that anything of value ever resulted from this work.

Like the Germans, scientists at the Mayo Clinic were also diversifying their approach to aviation problems. While Kendall worked furiously on the steroid angle, other Mayo scientists studied different ways to improve pilot performance. Starting in 1937 Mayo began its own sophisticated aviation research program.[24] The clinic's interest in aviation was not surprising; the Mayo brothers believed so strongly in aviation as the key to the future of Rochester that in 1929 they invested their own money to build the Rochester community airport. The clinic's close proximity to Minneapolis–St. Paul—the headquarters of Northwest Orient Airlines—facilitated collaboration between the rapidly growing medical clinic and the rapidly growing airline.

The Mayo Aero Medical Unit (which is what the wartime aerospace research unit was originally called) was, like the German government, interested in studying high-altitude flight—and like the Nazis at Dachau, Mayo also built a hypobaric chamber that could be used to simulate atmospheric conditions at high altitudes. The clinic's low-pressure simulation chamber debuted on March 26, 1939; it was used in a far more scientific—and humane—way than its Nazi counterpart. Mayo researchers immediately focused on the development of a small pressurized bottle containing breathable oxygen that could be used by pilots who were forced to parachute from their aircraft at high altitudes. This device, called the "bail-out bottle," was first tested in the hypobaric simulator and later by volunteer pilots who made high-altitude jumps using the new invention.[25] One of the test subjects participating in September 1942 was none other than Minnesotan Charles A. Lindbergh,[26] a good friend of the Mayo brothers. The bail-out bottle was proven effective by Lindbergh and others; it was rapidly incorporated into military operations, where it saved innumerable lives.[27]

High altitude wasn't the only stress affecting combat pilots. An even more important problem was g-force, the pressure that crushed a pilot's body during sharp turns at high speeds. Fighter planes of the 1940s were capable of flying faster and turning tighter than were the pilots who flew them; high-speed turns occasionally resulted in loss of consciousness. If cortin, the "great *flight* hope" of aviation, wasn't going to be available in the foreseeable future, could anything else be done to fight the stress of high g-force? Engineers at Mayo believed there was.[28] They constructed a human centrifuge that simulated the pressures and forces of high-speed combat flight. The huge machine, housed in its own building, spun the test subject (who sat in a small car attached to the end of a long rotor) in a wide circle. By 1942 the centrifuge was fully operational. Now that g-force could be studied, methods to overcome it could be developed and tested. Two advances merit special attention.

The first advance was developed by the project's leader, Earl Wood.[29] In talking with pilots he had discovered that yelling during high-speed turns somehow allowed them to tolerate extra g-force. Reasoning that if yelling was good, grunting might be better, Wood taught experimental subjects how to grunt (later known as the "M-1 maneuver"—"M" for Mayo) in the centrifuge and discovered that this action would, indeed, allow them to tolerate more g-force without losing consciousness. He quickly worked out the physiological mechanism behind this (it involves compression of the blood vessels in the belly and chest during the grunt, which forces blood into the head and thus prevents unconsciousness). The M-1 maneuver is still taught to military pilots as a way to enhance their tolerance of g-forces.

The second advance was even more significant. It was inspired by a basic physiological question: if contracting abdominal muscles (as occurs with a yell or grunt) shifted a little blood to the brain, would more compression shift even more blood? Frederick Banting (of insulin discovery fame) certainly thought so. Working with Wilbur Franks at the University of Toronto,[30] he developed a water-filled suit that compressed the arms and legs during turns and thus directed blood toward the brain.[31] This creation, called the "Franks Flying Suit," was produced by the British (approximately 800 suits were made) and tested in North Africa during 1942. Pilots hated the suits. The cuffs were filled with water; this made them distressingly hot and heavy.[32] The sloshing sensation they created gave the pilot an eerie sense of "floating," and the subsequent disorientation was almost as bad as the g-force effect itself. Using the human centrifuge, Earl Wood and his Mayo colleagues developed a different system utilizing five air cuffs that automatically inflated as the pilot began to turn the plane.[33] By squeezing the trunk and limbs at the right moment, the inflatable cuffs maintained blood perfusion to the brain—and kept the pilot conscious.

But would something that worked in a centrifuge simulator also work in real life? Scientists at Mayo bravely answered this question. The army allowed Dr. Wood to borrow a Douglas A-24 Dauntless dive-bomber, which he promptly renamed the "G-whiz."[34] Using himself and other Mayo scientists as test subjects (this had also been the case for experimentation within the centrifuge), he determined that sharp turns and high-powered dives were better tolerated using his new pneumatic "Anti-G suit."[35]

The Mayo Clinic wasn't the only place where research to support the war effort was taking place. In nearby Chicago, Enrico Fermi (winner of the 1938 Nobel Prize in Physics) had discovered how to sustain a nuclear chain reaction. As a result, his "Manhattan Project" was now in full swing.

Fermi would soon get some extra help.

CHAPTER 15

Plants, Politicians, and More Pessimism

What you do in this world is a matter of no consequence.
The question is, what can you make people believe that you have done?

—SHERLOCK HOLMES, *A STUDY IN SCARLET*

PERCY JULIAN WAS BECOMING TO THE SOYBEAN WHAT GEORGE WASH-ington Carver was to the peanut. Julian had recently discovered a protein in soybeans that could be used to coat paper and make it less flammable. His superiors at the Glidden Company passed some of this soy protein along to a Pennsylvania laboratory; the lab used it to create a fire-retardant product called Aero-Foam. The foam, which could be packaged in canisters and sprayed like shaving cream over oil or gas fires, was effective in extinguishing otherwise uncontrollable blazes, especially those occurring on ships at sea. Estimates of the number of sailors saved from death or injury by Julian's "Bean Soup," as it was affectionately called, run into the thousands.[1]

Russell Marker was also trying to put plants to good use. But he wasn't thinking about converting soybean protein to fire-retardant foam. Marker was hoping to perform a much different act of alchemy—the conversion of vegetables into money. The chemist returned to his laboratory at Penn State and immediately went to work.

As expected, the yams smuggled out of Mexico produced large quantities of diosgenin. Using a four-step version of his recently discovered degradation process, Marker converted all of the diosgenin into progesterone. The efficiency was impressive and the yield substantial. Yams were worth next to nothing. Diosgenin was cheap. But progesterone, which sold for $80 a gram, was worth more than its weight in gold.[2] Marker had discovered a biological gold mine.

Marker contacted his benefactors at Parke-Davis and explained what he had just accomplished. He expected to be met with unbridled capitalist enthusiasm. He wasn't.[3] The company declined to pursue the manufacture of progesterone, citing "the riskiness" of the venture.[4]

Rejected by Parke-Davis, Marker tried other companies. But the Parke-Davis rejection proved to be the kiss of death. "After I was convinced that Parke-Davis would not go into it, I tried other companies to get support. For instance, I tried Merck and they said that since Parke-Davis turned me down they would not go into it." He eventually realized that the only way his progesterone-synthesizing process was going to become a commercial reality was if he did it himself. Marker was as serious as cancer about this venture. In the middle of 1942 he resigned from Penn State, withdrew all of his savings from the bank—and got an indefinite separation from his wife.[5] Moving from the cool shadow of Mount Nittany to the sweltering tropical sun of Veracruz, Mexico, Marker went into business for himself.

Operating from his newly established base in Mexico, Marker conducted regular expeditions into the nearby jungle, where he found and harvested nearly ten tons of the wild yams he needed. Striking a deal with a local coffee merchant, he had the roots sliced—like cheap, greasy potato chips—sun dried, and packaged. He took these "yam chips" to Mexico City, where another food processor ground the chips like coffee. Using a unique chemical process of his own invention, Marker dissolved the ground yams in alcohol and evaporated the concoction into dense syrup. The syrup, which contained a concentrated form of the essential diosgenin precursor, was packaged in containers and shipped back to the United States.[6]

A new problem arose. Having left Penn State, Marker no longer had access to a chemistry laboratory. According to legend, he "borrowed" an American laboratory facility from one of his East Coast friends.[7] Employing his proprietary four-step chemical transformation, Marker quickly turned his yam syrup into seven pounds of waxy, amorphous progesterone. It was the largest batch of the female hormone that had ever been produced, and in 1943 its market value was a stunning quarter of a million dollars.

Marker had made progesterone; now he needed to sell it. The chemist knew nothing about business, but that wasn't going to stop him. Returning to Mexico City with a suitcase full of progesterone, Marker simply went to a public phone book and looked up "*laboratorios*." He found an entry for Laboratorios Hormona. It sounded promising, so he called and spoke to the lab's owner, Emerik Somolo, a Hungarian businessman and lawyer who had moved to Mexico in 1928 to make European-style pharmaceuticals. Somolo was skeptical of Marker's over-the-phone pitch, but he reluctantly arranged a meeting with the American. Somolo brought along his partner, a German medical doctor named Frederik Lehmann, who had joined him in 1933. Marker showed up at the meeting with a couple of loosely wrapped bundles in his arms. Imagine Somolo's and Lehmann's shock when the eccentric chemist opened them and revealed almost half the world's supply of progesterone! After Marker presented a rambling overview of his progesterone-synthesizing technology, Somolo and Lehmann offered him a partnership on the spot. It apparently didn't matter to them whether they understood the conversion process—as long as Marker could manufacture progesterone, they could sell it.[8] The three men formed a new company called Syntex (for "synthesis" and "Mexico").[9] Marker, who supplied the chemistry expertise—and tossed in his football-size blob of progesterone as part of the deal—received 40 percent of the stock. Inexpensive progesterone was about to become a reality.[10]

As the 1940s dawned, John Kennedy's stomach problems seemed to be improving. Maybe the strange steroid he'd taken for years, still known as DOCA, was finally helping? He continued having "trouble digesting his food" and "looked jaundiced—yellow as saffron and as thin as a rake,"[11] but at least his GI symptoms weren't getting worse. Unfortunately, other health problems were developing. In particular, his lower back was giving the future president problems. It first became noticeable in 1938, when declassified medical reports suggest JFK began to have deep, crippling pain in the right sacroiliac joint. Things took a dramatic turn for the worse in 1940 when he experienced spasms in his lower back while playing tennis; he described his subsequent agony as feeling like "something had slipped." The twenty-three-year-old was hospitalized at Boston's Lahey Clinic for ten days, during which time he began wearing a rigid back support brace.[12] The acute disability passed, but for the rest of his life Kennedy would have periodic attacks of intense, debilitating back pain. This would generate a recurring question: what role, if any, did his medicines—especially the steroids—play in weakening his bones or causing his back pain?

Kennedy returned to Mayo in 1940 for a checkup, and during his visit the young author hawked his new book, *Why England Slept*, in an interview on KROC radio—the station owned by Philip Hench's wealthy in-laws.[13]

The book, a slight reworking of Kennedy's Harvard thesis, received decidedly mixed reviews. One critic stated that "*Why England Slept* is not overwhelmingly brilliant."[14] A few better (and several worse) reviews can be found. Regardless of its quality, the book quickly became a best seller, although it's been argued that this occurred because his father bought thousands of copies. No matter. *Why England Slept* was not going to be the only successful book Kennedy would write.

Kennedy's back problems surfaced again in 1943 when injuries from his heroics on PT-109 (he was skippering the boat when it collided with a Japanese destroyer) pushed his physical issues back into the limelight. Kennedy was apparently seen again at Mayo,[15] where his growing list of medical problems was reassessed once more.[16]

Kennedy's next big health problem would occur in 1947; this time the future president would be visiting in England when crisis struck.

In 1942 young Kennedy was a junior grade lieutenant in the U.S. Navy. Philip Hench's military rank also included the word "lieutenant," but in his case it was lieutenant colonel. He held the title of chief of the Medical Service and director of the Army's Rheumatism Center at the Army and Navy General Hospital in Hot Springs, Arkansas.[17] As proof that high rank does not ensure happiness, Hench's son reported that his father "hated" his time in Hot Springs. It must have seemed a small conciliation when he was promoted to full colonel just before leaving the army. Hench's military years were a painful hiatus for a man intent on finding a cure for arthritis—and on discovering the truth about Walter Reed. As he had written to a friend on July 2, 1942, while preparing to fulfill the duties of his enlistment, "I am about to enter the Army and I am afraid that my study of the story of yellow fever will have to be indefinitely postponed."[18] True to his word, Hench's work on the subject had been largely abandoned for the past few years.[19]

Kendall had his own issues to deal with during the war. As a key member of the National Research Council (NRC) working on the development of "cortin" and its potential military applications, he had a well-defined mission in front of him. The goal was to make compound E, which Kendall and others now believed was the most potent adrenal steroid in terms of its ability to affect muscle function, tolerance to stress, and many other physiological characteristics. But from a chemist's perspective, compound E was going to be an extremely difficult material to make. Kendall pushed the committee to synthesize compound A first—chemically, this would be much simpler, and it would facilitate the subsequent synthesis of compound E.

But there was a problem with the plan. Even though compound A was less complicated to make than compound E, progress on the synthesis of compound A was still very slow. Kendall had plenty of excuses for the poor showing: he blamed war-related attrition (many of the scientists who could have been working on this problem were diverted by the war into other areas), material shortages, and other setbacks. Dwight Ingle, his assistant, saw it differently; he attributed the lack of progress to Kendall's penchant for constantly doing things the hard way. Although he wasn't blessed with a lot of personal insight, Kendall appears to have recognized his own tendency to shoot for lofty goals when simpler ones might be more practical. As he once told an associate, "I want to grow a great big oak tree. I am not interested in a bunch of blackberry bushes."[20] Big projects, even if they were impractical or unfeasible, captivated Kendall's attention more than the realistic endeavors he could accomplish in a reasonable time frame. The project to make compound A, and then parlay this into the manufacture of compound E, was exactly the type of impossible mission that captured Kendall's interest.

Predictably, it wasn't Kendall who finally manufactured compound A. In 1943 Tadeus Reichstein synthesized the elusive steroid starting from bile acid obtained from oxen. The extremely low efficiency of the process—more than 10,000 pounds of ox bile were needed to make 1 pound of compound A—made this method far too expensive to be of any potential commercial practicality. But now that Reichstein had beaten him in the race to make compound A, there was little doubt in Kendall's mind that the Swiss chemist would also beat him to the final goal—the synthesis of compound E.[21]

Reichstein, working in a small laboratory in war-beleaguered Europe, solved a problem that Kendall and his colleagues had been unable to surmount—even though Kendall's group had the vast resources of the United States backing them. Now that the war was winding down, the importance of compound E was falling. And although Kendall was probably unaware of it, his personal reputation—his "stock" as a scientist—was also falling. Of all the NRC goals established at the onset of the war, Kendall's project to make compound E was the only one that didn't succeed as expected. In 1944 all members of the NRC committee except Mayo and Merck & Company quit working on the synthesis of compound E. With the synthesis of compound A hopelessly inefficient, and the rationale for pursuing the synthesis of compound E becoming increasingly more obscure, the future of adrenal corticosteroids never looked dimmer.

Halfway across the country, another person of interest was contributing to the war effort at this time. Luis Alvarez had just been asked to join the Manhattan Project.

His assignment? To help build an atomic bomb.

CHAPTER 16

Good-bye Marker,
Hello Sarett

> *You know that a conjurer gets no credit when once he has explained his trick; and if I show you too much of my method of working, you will come to the conclusion that I am a very ordinary individual after all.*
>
> —SHERLOCK HOLMES, *A STUDY IN SCARLET*

THE WORLD WAS STILL AT WAR IN EARLY 1945. BUT NOT FOR LONG.

On July 16, 1945, Luis Alvarez witnessed his latest contribution to the war effort from the seat of a B-29.[1] At an altitude of 30,000 feet and a distance of approximately fifteen miles, he watched as the detonator he had designed triggered the first nuclear explosion at the Trinity Site in Alamogordo, New Mexico. Alvarez could not find words to describe the scene as the first atomic bomb went off. Instead, he sketched what he saw. His illustration of a mushroom cloud, drawn as he circled the blast, is the first ever made of this now-famous icon. But it was not the last mushroom cloud Luis Alvarez would see. Three weeks later he'd fly in a B-29 again, this time trailing the *Enola Gay* as it passed over Hiroshima and dropped its nuclear device. Three days after that the "Fat Man" plutonium bomb—equipped with Luis Alvarez's detonator—would fall on Nagasaki and end the war with Japan.

Russell Marker was also "at war," and one of his adversaries was German. But the other wasn't Japanese. He was a Hungarian.

Marker was fighting with his partners at the newly created Syntex Corporation.[2] As Luis Alvarez was putting the final touches on the "Fat Man" atomic bomb prototype, Marker was looking for profits from his new steroid-manufacturing company—and there weren't any. They had been reinvested in order to grow the business. Marker, never the most trusting person, immediately assumed the worst—that his failure to receive any money meant his partners were either cheaters or incompetent. Over the next several months he pulled out of the company, taking with him anything of value he could get his hands on. But the most valuable thing he took with him wasn't in the lab; it was in his head. Marker was the only one who knew how to chemically convert yams into progesterone, and he had deliberately kept the secret of this process to himself. There were no written instructions describing the steps involved, and rather than label the various reagents properly, he'd put cryptic codes on the bottles.[3] When Russell Marker walked out of Syntex, the company essentially shut down.

His ex-partners, Emerik Somlo and Frederik Lehmann, needed immediate help to restart the business. Mexico was not yet a place where chemists could be easily found; indeed, some have claimed that as of 1945, no Mexican had yet obtained a PhD in chemistry. Fortunately, an associate returning from Cuba told Somlo and Lehmann about another Hungarian, George Rosenkranz, who was currently working in Havana. Fearing Nazi persecution, Rosenkranz had moved first to Zurich and then on to Havana, where he took a job as director of research for the country's largest pharmaceutical manufacturer.[4] Rosenkranz was a well-known steroid expert. Somlo and Lehmann approached him about joining Syntex. They interviewed him in Mexico City on August 6, 1945—the day the bomb fell on Hiroshima. Something about the vague, potentially risky offer intrigued Rosenkranz, and he accepted the position. Rosenkranz quickly figured out what Marker had been doing (he had been studying Marker's publications for years and was already familiar with this work), and Syntex was back in business. In a big way.

Marker was also quickly back in business. He moved to Texcoco, just outside of Mexico City, and created a new company called Botanica-mex. He had managed to produce only 1 kilogram of progesterone during his time with Syntex, but within a few months at Botanica-mex he made several kilograms of the steroid hormone (competition was bringing down the price of the hormone, but even at $50 a gram his progesterone stash was worth hundreds of thousands of dollars).[5]

Once again, the work did not go smoothly for Marker. "Unidentified outsiders" physically harassed his workers, creating labor unrest.[6] In March 1946 Marker decided he didn't need the headache of management anymore. Botanica-mex ceased to manufacture progesterone, and the assets were sold to Gedeon Richter Ltd., a growing Hungarian drug manufacturing entity. A new company, called Hormonosynth, was formed in Mexico City, and production was restarted.[7] Marker, now an employee of the new company, discovered an even better yam (called "barbasco") that contained five times as much diosgenin as the "*cabeza de negro*" that he had been using until now. Somewhat predictably, Marker once again grew tired of his situation. In 1949, citing "personal reasons," the forty-seven-year-old chemist retired from Hormonosynth and permanently gave up chemical research. Without Marker, the company once again underwent significant corporate reorganization, and the resulting new commercial entity became part of Diosynth.

Russell Marker's contributions to the steroid story in general, and the cortisone saga in particular, end here. Marker had found a way to manufacture progesterone in a relatively inexpensive way.[8] Progesterone would ultimately become an important starting material for the synthesis of cortisone. Without Marker's contribution, the eventual outcome of the cortisone story would have surely been much different.[9]

While Russell Marker was departing from the cortisone story, a fresh new face was arriving. His name was Lewis H. Sarett—a young, handsome chemist of extraordinary ability. And he was about to play a role in the cortisone story that would nearly rival that of Reichstein, Kendall, and Hench.

Lewis Sarett was born in December 1917 in Champaign, Illinois.[10] His father, an outdoor enthusiast and poet, became a professor at Northwestern University. Sarett attended high school in Highland Park, Illinois, a short drive from the spot in Oak Park where Hemingway was born and an even shorter drive from the University of Chicago, the place where Luis Alvarez learned his physics and where Enrico Fermi handed the Manhattan Project the first self-sustaining nuclear chain reaction. "Lew" was recognized as a gifted thinker early in his life. He taught himself chess and eventually became an excellent player (and one of the first to play regularly against a computer). Although he had a flair for math, a beginner's chemistry set launched him on the road to his career. He graduated Phi Beta Kappa from Northwestern University in 1939 with a degree in chemistry.

Sarett continued his studies at Princeton, where he became interested in organic compounds. When the war began heating up, so did interest in steroids. Sarett's graduate studies soon focused on the search for simpler methods of steroid synthesis. After graduation,[11] Sarett headed off to work with Merck, choosing that company specifically because it was a major player in the 1942 "cortin consortium" sponsored by the National Research Council. In an effort to bring Sarett up to speed in state-of-the-art cortin research, Merck asked for a favor from the Mayo Clinic. It was granted, and in early 1942 Lewis Sarett came to Kendall's laboratory for three months of intense training and introductory research.

Sarett made some remarkable accomplishments during his short time at the Mayo Clinic, including the identification of a key intermediary in the synthetic pathway for manufacturing compound E. Perhaps more important, the young man impressed his mentor, Dr. Kendall.[12] The two would become friends as well as colleagues, and the synergy started at Mayo would extend for years and span beyond the completion of the cortin project. When Sarett left Kendall's lab he moved to Merck headquarters in New Jersey; Kendall visited him there frequently. The meetings inevitably focused on two topics: the synthesis of compound E and chess. Sarett's love of the game equaled Kendall's, and the two would play whenever they were together. Even when they were apart they would play by mail.

The synthesis of compound E and the game of chess are similar. Multiple moves, complex rules, strategy, and imagination are essential to both. Now both scientists were ten moves into their latest friendly, but highly competitive, game—the contest to synthesize compound E.

Lew Sarett married Mary Adams Barrie on March 1, 1944, and went on to have two daughters. Unlike the somewhat unidimensional Dr. Kendall, Sarett achieved substantial balance in his life. He became a scratch golfer. Spurning the rigidness of academics, he decided to enter the business world.

> Let's start with why I joined an industrial research organization rather than academia. The answer is embarrassingly simple. First, my father was a professor and I wasn't thrilled with the idea of becoming a "me-too" teacher. Second, the fathers of most of my high school friends had commercial jobs as bankers, salesmen, and executives. They all seemed to be having a great time of life, whereas I felt that the academics weren't having much fun; most seemed to take themselves too seriously.[13]

Sarett's marriage to Mary Barrie ended in divorce. On June 28, 1969, he married Pamela Thorp, a microbiologist at Merck. They had a son and daughter of their own. When Lew Sarett finally decided to retire in 1982, the couple built a home in Viola, Idaho, where they could spend time outdoors, and where Sarett could enjoy two more of his multiple interests. Hunting and fishing.

It turns out that Sarett wasn't the only hunter to retire in Idaho.

CHAPTER 17

Hench Returns to Mayo

It has long been an axiom of mine that the little things are infinitely the most important.

—SHERLOCK HOLMES, *A CASE OF IDENTITY*

WORLD WAR II WAS OVER; LIFE WAS BEGINNING TO RETURN TO NOR-mal; and, just like his idol Sherlock Holmes, Philip Hench was beginning to care once again for the little things. Colonel Hench left the army in 1946, although he remained a consultant to the army surgeon general for matters relating to rheumatology.[1] Now it was time to get back to his real work and interests. Compound E was still unavailable, so Hench pursued the yellow fever story, political issues, and his ongoing rheumatology research.

Most of the activity involving yellow fever centered on Hench's copious correspondence. He was trying to identify a potential publisher for his eventual treatise on yellow fever, and Schuman's Publishing in New York was showing interest. Not even a case of winter flu could keep Hench from pursuing a publishing lead for the book he planned to write. On January 26, 1946, he followed up with Henry Schuman, the company's president:

I have your letter of December 8, and regret this delay in answering.

I have been in the Army three years and have been traveling around rather extensively the last three months. I was inactivated from Washington just before Christmas but caught the prevailing respiratory infection and I am just now back at work.

Hench had laid out his proposal for the yellow fever book well enough to grab Schuman's interest. "I appreciate indeed your interest in the study that I am making on Walter Reed and the whole yellow fever episode. I think you are right in saying that I am trying to do the definitive job on the subject. I have about 25 or 30 files actually on hand and have collected a surprising amount of original data, including original records of all sorts."[2]

But there was a problem. Despite interest from a publisher, Hench knew the story would take a long time to finish. He was severely overcommitted, especially with regard to his academic pursuits. "However, this has been in the nature of a hobby and must come second to two or three other jobs—first, my practice at medicine here, second, my annual publication of the Rheumatism Review, and third, my clinical research which, despite my intense interest in the yellow fever story, I must continue, and which fortunately, is getting a little 'warmer' all the time."[3]

Hench's discussions with Kendall in 1941 regarding treatment options for rheumatoid arthritis may have resulted in a decision to try compound E when it became available, but most of their talks involved other matters. Hench was hoping to relieve arthritis symptoms by giving his patients substances that would cause them to become jaundiced.[4] He assumed that the mysterious "substance X" must be increased during jaundice, and that this explained why jaundice seemed to help. The problem? Jaundice was hard to produce safely, and the substances one could give a patient to induce it were usually poisonous, especially to the liver. Many of Hench's conversations with Kendall involved his desire to understand the nuances of these agents and the biochemical changes they produced.

As Hench was exiting the army, a new jaundice-producing agent was being studied in Sweden.[5] Called lactophenin, it was a moderate-to-mild hepatotoxin that safely produced jaundice in 40 to 50 percent of the patients who took it. Preliminary information suggested that those who became jaundiced experienced relief from their rheumatoid symptoms. Lactophenin was an avenue of research that Philip Hench naturally wanted to explore.

The yellow fever story was also beginning to have political implications for Hench. His efforts to glean information about Reed and his work, especially from Cuban sources, were frequently complicated by the bad feelings many Cubans harbored over the American interpretation of past events. Specifically, many Cubans believed that one of their own countrymen, Carlos Juan Finlay, was not being given the credit by America that he deserved.

Carlos Finlay, a balding, bespeckled man with true sideburns,[6] was born in Cuba in 1833, the son of a Scottish physician.[7] He was educated in France, and later received a medical degree from Jefferson Medical College in Philadelphia (where he studied under a protégé of Dr. Brown-Séquard). He returned to Cuba and practiced medicine. Over the following decades his experiences and observations led him to conclude that yellow fever was somehow transmitted by mosquitoes.[8] Although he was unable to prove it definitively with the limited scientific resources available to him, his observation formed the basis for Walter Reed's subsequent studies. Many Cubans indignantly believed that Reed got all the credit for what was essentially Finlay's idea.

Hench was keenly aware of this uncomfortable situation. As a true fan of Walter Reed, he adamantly believed that the American army major was the main reason for the success of the yellow fever investigations, and that he therefore deserved most of the credit for the discovery; on the other hand, Hench was sensitive to Finlay's role and the concerns of the Cubans. Writing to a friend on April 16, 1946, he noted, "I agree with you that Finlay has not received as much credit in this country as he deserves but the Americans have certainly not received the credit they deserve in Cuba. I have visited there many times and the misunderstanding is painful. I hope by my own writings to do what I can to heal the wounds."[9] Such was the nature of Philip Hench—if he wasn't trying to heal wounds caused by rheumatological diseases, he was trying to heal those caused by bad politics and the shortcomings of human nature.

By 1947 Hench had a brewing Cuban crisis of sorts on his hands. Of course, the more famous—and potentially far more serious—"Cuban missile crisis" wouldn't occur for another fifteen years. Coincidentally, at this very moment one of the men destined to play a pivotal role in 1962's biggest political event didn't look like he'd live long enough to fulfill his destiny. John F. Kennedy was dying.

During 1946 Kennedy ran for and won a seat in the U.S. House of Representatives. The race seemed to take an unusually severe physical toll on him. According to some, "He was actually like a skeleton, thin and drawn."[10] He suffered from worsening abdominal pain, fatigue, nausea, and vomiting. "People around him noticed his bulging eyes and jaundiced complexion."[11] Those closest to him expected some improvement after his victory. It didn't happen. The future president continued his slow physical deterioration.

In 1947, while visiting in England, Kennedy experienced a severe exacerbation of his illness and sought medical help.[12] He was hospitalized in London, the same city where Addison had practiced medicine a century earlier. Kennedy underwent intensive testing, and a shocking diagnosis was made.

In a cruel twist of irony, it was determined that JFK had developed Addison's disease.[13] Like Jane Austen a century earlier, his adrenal glands were impaired, and his cortin function was deficient.

Why? How? Most of Addison's original cases were caused by tuberculosis. Kennedy almost certainly did not have tuberculosis, a fact that gave his brother Robert the ability to claim to the media—without technically lying—that JFK did not have "classic" Addison's disease. More likely, Kennedy's condition was a form of "secondary Addison's disease" that can be seen in patients who take steroids for chronic problems. In this condition the adrenal glands become suppressed by the extra steroids, and over time the glands shrink and may permanently lose their ability to make cortin.

In JFK's case, adrenal suppression was probably caused by years of using DOCA and other cortinlike steroids to treat his colitis. As long as he took a sufficiently large daily dose of "steroids," his Addison's disease was masked or controlled. But if he stopped taking his steroids, even for a short while, his own suppressed adrenal glands could no longer manufacture sufficient cortin to support him. Unable to make his own cortin when he needed it, JFK would quickly develop the symptoms of full-blown Addison's disease. The greater the physical or emotional stress Kennedy was under, the more severe his condition became.

JFK's situation in London appeared life threatening. The attending hospital physician told Kennedy's friend Pamela Churchill, daughter-in-law of Winston Churchill, "That young American friend of yours, he hasn't got a year to live."[14] The doctor was almost right. Kennedy recovered enough to leave the hospital and attempt the trip back to America aboard the *Queen Mary*, but en route he relapsed and became so sick that a priest administered the last rites to him in anticipation of his death at sea. Kennedy recovered enough to survive the trip, but he was so weak upon his arrival in New York that he was carried off the ship on a stretcher. For the next few years his health would wax and wane unpredictably.

The worst lay ahead for JFK. The future leader of the Free World would soon face another adrenal-related life-threatening health crisis.

Now that the war was over, Hench found he had time to deal with the little things. Two in particular. First, he wanted to catch up on the medical meetings of national importance that he'd missed while in the service. "I expect to be in New York May 24 and 25 and then will go to the Atlantic City meetings. Having been in the Army for almost four years I haven't had an opportunity of attending a meeting of the Association of American Physicians since I was elected (as a Fellow of the Association)."[15]

Second, he wanted a little vacation time. Hench had missed spending time with his wife while he was in the army, and—helped by the Kahler family money—travel with her was now a high priority. "Last month I went to the American Medical Association meeting, stayed at the Mark Hopkins for a week and then went to . . . Santa Barbara for five days rest with Mrs. Hench."[16] Mixing business with pleasure seemed the rule for the reunited Henches in 1946. In August Hench wrote to a friend, "I now have the exact schedule for the meeting in Oklahoma City, which I am attending. . . . Mrs. Hench and I have decided to go from Oklahoma City to Colorado Springs for a week or ten days of rest."[17] Hench's first year out of the army involved intense business and recreational travel, even by Hench's legendary standards.

Hench was promoted to full professor of medicine in 1947. On the same day he received his promotion he also received a scathing note from the Mayo Clinic medical library demanding that he return an overdue copy of the *Southern Medical Journal*.[18] Hench had recently grown accustomed to the respect that came with military rank, but academic rank means nothing to an angry head librarian. More important, his promotion within the clinic did nothing to reduce the constant, unrelenting stress gnawing away at him. Nineteen forty-seven seemed to bring many of the same headaches for Hench as did 1946. While his duties and responsibilities at Mayo demanded his attention, his desire to work on the yellow fever story was becoming impossible to ignore much longer. He wrote in October 1947, "I realize that I have a real responsibility toward the yellow fever story. The only trouble is that I have so many other pressing responsibilities in the field of rheumatology. Nevertheless, I am slowly keeping at work in odd hours and if my health lasts, the job will be finished."[19]

"If my health lasts"? Was Philip Hench developing health problems? If so, those around him seemed unaware of it. One of his good friends, a Frenchman, knew Hench well during this period of his life and did not describe him as a man with health problems.[20]

> At work and during discussions, Hench was attentive, had a positive attitude and was enthusiastic in defending his ideas but discussed with indulgence and

sympathy those of others. Outside his work, Hench was not the austere and pensive scholar. He was a gay fellow, vivacious and noisy, who loved travel, adored music. . . . While with a patient he would bend over with amazing simplicity, searching deeply for the human contact; he questioned then examined with great warmth . . . for besides having the mind of brilliant clinician, he also had the soul of a knight of science.[21]

If his health concerned him, it didn't show. Dr. Hench seemed at peace with life. Unfortunately, Dr. Kendall did not.

Kendall's oldest son, Roy, died of a blood cancer while training as a medical resident. Kendall was in Atlantic City attending a meeting with some of his colleagues when he got the news. Many of those around him were "stunned and unable to find words to express [their] feelings."[22] In contrast, Kendall, who was somewhat older than most of his companions, "spoke quietly of the cruelty of disease and that man must work toward the reduction of it. Nothing short of his own death would prevent his striving toward that goal."[23] Kendall comforted those around him rather than seeking his own comfort in this time of need. The tragic situation was repeated shortly afterward when another of Kendall's sons, Norman, committed suicide not long after being discharged from the military. Again, Kendall seemingly kept his focus, although the effect on his wife, Rebecca, was more predictable. She suffered a "nervous breakdown" and required hospitalization.[24] Years earlier she had begun experiencing what has been described as "a series of periodic mental illnesses."[25] It appears that the circumstances in the Kendall household aggravated a depression that would intermittently cripple her for years.

It is difficult to assess the impact of Kendall's family situation on his personality—or of his personality on his family life. As was the custom in Rochester, no one spoke of his wife's malaise. There is almost no mention of Kendall's children in his autobiography. Although he discusses his spouse, parents, siblings, aunts, uncles, grandparents, and others, information about his children seems eerily absent. One can only assume—or at least hope— that his omissions reflected the psychosocial effects of so many awful family events on Kendall's recollection of life during these daunting times.

CHAPTER 18

Push On? Give Up?

All the cards are at present against us.

—SHERLOCK HOLMES, *THE PROBLEM OF THOR BRIDGE*

THE YEARS 1946 TO 1948 WERE QUITE POSSIBLY THE WORST IN Edward Kendall's entire life. Personally, he'd suffered the tragic loss of two grown sons. Professionally, his research on cortin was going down in flames.

The "war committee" of the National Research Council (NRC) had made adrenal cortical research its number one priority, and, as noted previously, the project hadn't exactly gone according to plan.[1] They were trying to produce compound A in the belief that it would be easier to make this substance than its biochemical big brother, compound E. But even making compound A wasn't easy. Working with Kendall, scientists from Merck & Company adopted Reichstein's new technique and converted hundreds of pounds of ox bile (the preferred starting material) into just under 100 grams of compound A. It wasn't much, but with this amount of the mysterious steroid on hand it was now possible to conduct meaningful experiments on its biological activity.

Unfortunately, the results of these early tests with compound A were disappointing. By itself, compound A couldn't keep adrenalectomized animals alive—the addition of other substances derived from the adrenal cortex was necessary. Compound A definitely wasn't *the* cortin investigators were searching for. But it clearly exerted profound effects on muscle and metabolism. Dwight Ingle, using his rat model, had already shown that compound A augmented muscle performance. Others had demonstrated its profound influence on carbohydrate metabolism, including its ability to promote glycogen deposition in the livers of adrenalectomized rats—presumptive evidence of "anabolic" activity. And like other anabolic-type steroids, it causes glands like the thymus and adrenals to shrink when given to healthy animals.

Based on promising preliminary results, members of the NRC group conducted a series of experiments and clinical trials using compound A. In March 1946, at the Atlantic City meeting of the Federation of American Societies for Experimental Biology, a session was convened to review their findings. Hopes were running high. Tadeus Reichstein, the steroid wizard from Switzerland, was in the United States and attended the meeting. Also present was Randolph Major, director of research for Merck & Company; he was largely responsible for Merck's substantial—and very expensive— contributions to the adrenal cortical project. Given Merck's huge financial investment to date, he could not help but think his job was potentially in jeopardy at this moment. Positive findings would quickly cheer the room.

But no cheer was forthcoming. Investigator after investigator presented the results of his research, all with the same basic conclusion: compound A had little clinical usefulness.

Dr. Edwin Kepler of the Mayo Clinic stood up and gave a brief report of his recent findings: he had specifically tested compound A as a treatment for patients with Addison's disease and emphatically concluded that it had no value. Kepler had used massive doses of compound A (200 milligrams daily) to ensure that underdosing was not responsible for the negative result. The foul fog of despair filled the room. Kendall later noted that "as we walked out of the auditorium Dr. Konrad Dobriner of the Sloan-Kettering Foundation remarked to me, 'This meeting never should have been held. Now no pharmaceutical company can be persuaded to try to make compound E.'"[2]

Dobriner's position was logical enough. Compounds A and E were almost identical—they differed by just one atom of oxygen attached to carbon 17. How important could this minor deviation possibly be? Indeed, all preliminary testing to date suggested that the major differences between compounds A and E would be quantitative rather than qualitative. As Kendall put it, "Many of those in a position to have an opinion thought that E might be two or perhaps three times as active as A."[3] Unfortunately, you don't need to be a mathematician to quickly grasp that an activity of two or three times "zero" is still "zero." Compound A was turning out to be the *Hindenburg*, *Titanic*, and *Ishtar* of medical research—all rolled into one biochemically unimpressive molecule. It was impossible to envision a bigger, more high profile, and more expensive failure than the study of compound A—except for the now very real possibility that compound E might prove to be a similar disaster.

Kendall had multiple personal and professional reasons to be depressed. He had spent the last eighteen years of his life on the "quest for cortin," and at this point the project reeked of failure. And he was only four years away from the mandatory retirement age at the Mayo Clinic. Addison would have been leaping headfirst over the garden wall by now. But not Kendall. If he was depressed by tragedy or setbacks, it didn't show. Kendall merely suggested in his autobiography that "March 1946 was a low point in the development of compound E."[4]

Kendall returned to Rochester after the Atlantic City meeting, buried his son, and refocused his work. The clinic's laboratory had already prepared more than 30 grams of compound E by extraction from bovine adrenal glands—these were the spoils of the Parke-Davis and Wilson projects. Mayo investigators had tested small amounts of compound E on animals, and "generous" quantities had been given to other investigators. The results of these experiments were beginning to trickle back to Kendall. For reasons that are not totally clear, these preliminary results, unlike those involving compound A, inspired some optimism in the chemist. No one knows exactly what Kendall "saw" when he looked at the results from these experiments (and he was so frequently wrong in his interpretation of physiological data that his opinion on the subject hardly mattered), but he somehow concluded that "our results strongly indicated that there would be a place for (compound E) in clinical medicine."[5] Kendall, perhaps foolishly, was not giving up yet.

Nor, it turns out, was Merck & Company. A conservative business operation would have curtailed its research efforts long ago. Even the most entrepreneurial risk-takers would likely fold their tents at this point. But on October 24, 1946, Randolph Major held a Merck-sponsored meeting on the East Coast with Kendall and his associates. Amazingly, an interim decision was made to continue the project by attempting to manufacture a small amount (roughly 5 grams) of compound E, enough to explore the difficulties involved in its production. To lead this undertaking, all eyes turned toward young Lewis Sarett.

Lew Sarett had returned to Merck's research facilities in Rahway, New Jersey, after leaving the Mayo Clinic in 1942. As expected, his work now focused entirely on the attempt to synthesize compound E. He struck pay dirt in December 1944, when he successfully manufactured a few milligrams of compound E starting from ox bile—the same material from which Reichstein had first made compound A. Sarett's achievement proved for the first time that compound E could be created from precursors not already present in the adrenal gland.[6]

The first synthesis of compound E was a major chemistry accomplishment with clear-cut scientific significance; only twenty-six years old, Sarett had now earned himself a place in many yet-to-be written scientific history books. Unfortunately, from a manufacturing point of view, it was a slow process with extremely low yields (60 kilograms of starting material were needed to produce 1 gram of compound E). It was also a highly difficult synthetic process, involving temperatures as high as 300 degrees—and as low as 100 degrees below zero—Fahrenheit. The tiniest mistake, sometimes occurring after weeks of work, left the chemist holding a flask filled with worthless sludge. As a scientist, Sarett had accomplished something remarkable—the first synthesis of compound E. But commercially, the process he had invented was useless. Sarett needed to come up with a new method for the synthesis of compound E. He accomplished this feat in just under three years. The details are tedious: nearly forty chemical steps were involved.[7] The crucial breakthrough, for those interested in the nuances of chemistry, involved the discovery of a better way to add an oxygen atom to the carbon 17 molecule of the steroid ring.

Kendall himself helped Sarett solve another major dilemma in the final step of the conversion process—a problem that was only slightly less difficult than placing an obstinate oxygen atom on carbon 17. The dilemma? A double bond had to be inserted into one of the rings of the steroid carbon skeleton. Working with the best chemist in his lab, Kendall developed the "Mattox-Kendall procedure";[8] it allowed 1,000 grams of a crucial intermediate to be converted into 900 grams of compound E, which gave an amazingly high yield for such a complex chemical reaction.[9] A scientist from Merck & Company brought an ample supply of the essential precursor to Kendall's lab, and on Thanksgiving Day, 1948, Kendall and the chemist from Merck converted this batch of precursor into compound E, which was then sent back to Rahway. It was a historic day, and like the discovery of thyroxin thirty years earlier, it involved Kendall spending yet another holiday in his laboratory.

Because of the efforts of Sarett, Merck & Company, and to a lesser extent Kendall, it was now possible to synthesize compound E efficiently enough to manufacture it commercially—despite the fact that doing so now involved the world's most complex chemical synthesis.[10] Merck geared up its facility to meet the manufacturing challenge that lay ahead.

Sarett's success meant that compound E would soon be available in relatively large quantities. Now what? The proverbial cart had been placed before the horse. There was still no obvious use for compound E. It caused some relatively limited physiological effects when given to animals, but the impact of the drug was decidedly underwhelming. It did not make fighter pilots fly higher. It didn't help athletes run faster, jump farther, or lift more. Like compound A, it helped—but did not fully correct—Addison's disease. And unless someone could find a way to make it water soluble (so it could be taken by mouth instead of requiring a painful injection into muscle using a long, large-bore needle), it wasn't likely to be used at all in these patients; after all, adrenal cortical extracts (which contained a mixture of compounds in addition to compound E) were cheaper, could be given by mouth, and were much more effective at the treatment of Addison's disease.[11]

Charles Slocumb, a rheumatology partner of Hench's at Mayo, had for several years described compound E as an "unheralded treatment in search of disease." Later, that position was modified slightly, but prophetically. He said Kendall had "a substance for which he did not have a disease and Hench had a disease for which he did not have adequate treatment."[12]

On April 29, 1948, Merck hosted a meeting in New York City to reassess the situation with regard to research on compound E. Specifically, what studies should be undertaken next to assess its utility? Kendall was invited. Despite the never-ending series of setbacks, he seemed cautiously optimistic. Sarett's advances in synthetic technique meant that limited amounts of compound E were slowly becoming available to investigators. As Kendall put it, "now nothing blocked the way of the clinical study on an impressive scale." He saw this as an opportunity to finally fulfill his dream of completely evaluating compound E. "I entered that after-dinner conference with suppressed emotion and with hopeful expectation that this meeting would mark the end of the long road that we had traveled since 1930."[13]

Once again, the meeting didn't turn out as expected. Merck had now invested over $13 million in the development of compound E—and for what? According to Kendall, "as the evening progressed, the atmosphere seemed to turn from cool to frigid . . . whether it was fear of the use of a compound that had cost so much to prepare, or hesitancy to be associated with a clinical investigation that probably would result in failure was not evident."[14] No one attending the meeting was particularly interested in conducting any further studies with the newly available compound E. Dr. Randall G. Sprague of the Mayo Clinic reluctantly volunteered to try a small amount on a few of his patients with Addison's disease, but no one else came forward with a specific clinical proposal. The lack of enthusiasm spelled potential disaster. Randolph Major, the Merck official running the compound E project, had warned Kendall prior to the meeting that "unless some definite use for compound E could be found, he was not confident that Merck & Company would continue this project."[15]

In his autobiography Kendall makes no comment about the death of his son Roy in 1946, or the suicide of his son Norman in 1947. But with regard to this 1948 meeting involving his life's work, he had plenty of comment about his sadness. "No ray of light could pierce the gloom of that gathering. It reminded one of W. E. Henley's *Invictus*: '. . . the night that covers me, Black as the Pit from pole to pole.'"[16]

Nick Kendall may have had problems seeing things for what they really were, but this time his perceptions were probably correct. All the cards really were against him. "April 29, 1948, was for me the low point of the entire investigation of the adrenal cortex."[17]

CHAPTER 19

The Decision to Test Compound E on Rheumatoid Arthritis

I play the game for the game's own sake.

—SHERLOCK HOLMES,
THE ADVENTURE OF THE BRUCE-PARTINGTON PLANS

CHARLES SLOCUMB HAD BECOME THE MAYO CLINIC'S SECOND RHEU-matologist in 1931; Howard Polley joined up as the third in 1942.[1] The two junior men had faithfully supported Phil Hench's research into new ways to treat rheumatoid arthritis. The latest effort involved the use of lactophenin, the liver toxin from Sweden that caused jaundice in many patients and relieved arthritis symptoms in some.[2] These reports had stimulated Hench's interest, and the three investigators were now attempting to confirm and repeat the Scandinavian observations at the Mayo Clinic.

Many patients came to Rochester hoping to get rid of jaundice. But in July 1948 two patients arrived in the strange hope of *becoming* jaundiced.[3]

The two had severe, long-standing rheumatoid arthritis. They came as volunteers hoping to receive lactophenin on the chance that it would somehow relieve their symptoms. The therapy was clearly experimental, but given the complete lack of alternative therapies, lactophenin offered at the very least a ray of hope for the two test subjects.

Both patients were accepted for study and admitted to Saint Marys Hospital, where treatment with lactophenin was initiated. The first patient responded as hoped: jaundice developed, and with it came a dramatic remission of arthritic symptoms. Unfortunately, the second patient, "Mrs. G." from Kokomo, Indiana, did not respond. Despite aggressive administration of the hepatotoxic drug, she stubbornly maintained a nonjaundiced milk-white complexion. Her arthritis raged unchecked, making it impossible for her to raise her arms over her head, lift a book, or, at times, even roll over in bed. After administering as much lactophenin as he could safely give to her with no success, Phil Hench suggested that Mrs. G. return to Kokomo "while we study the matter further."

Mrs. G. would have none of that suggestion. Only twenty-seven years old, she already had a reputation as a professional patient because of her participation in other clinical trials (including one conducted by the Eli Lilly Company back home in Indiana that had tested streptomycin as a potential antirheumatoid agent). After seeing the dramatic relief that jaundice produced in the other experimental patient, she was buoyed by unfounded optimism. Summoning her most defiant attitude, she told Hench "No." Mrs. G. said she had "come for relief and wasn't leaving until she got it." To borrow an expression used by doctors in training, she "grew roots" in her hospital bed, stubbornly refusing to go home until she had been a guinea pig for another, hopefully more successful experimental therapy.

Although stubborn, Mrs. G. was an extremely cooperative and likable patient. Rather than being put off by her demands, Hench seriously considered alternative ways to help her. For the next several weeks he discussed the situation with Drs. Slocumb, Polley, and others. Unfortunately, no one had any brilliant ideas.

Out of desperation more than anything else, Hench considered the possibility of using compound E to treat Mrs. G. Hench was aware that Dr. Randall Sprague, a Mayo Clinic endocrinologist, had in his possession 2 grams of compound E that he'd received from Merck & Company in order to study its usefulness in a patient with combined Addison's disease and diabetes. Hench wanted some of that sample for Mrs. G. He called Sprague and made his request. Years later Sprague would recall "being in the middle of hospital rounds at the time of the call on a busy August work day and conversing—but mostly listening—for about 45 minutes." Sprague not only declined to contribute some of his compound E, but tossed in the opinion that he thought Hench's plan was "an absurd idea." Sprague had no desire to waste his precious aliquot of compound E on an application he considered to be preposterous.

Hench remained undeterred. There was another potential route to compound E. He picked up the phone again and called his colleague Dr. Kendall. Hench knew Kendall also had some compound E on hand, although it was a smaller amount than Sprague's supply. Hench hoped to charm the chemist into giving him of some of it. Unfortunately, Kendall wasn't in the office. Hench left a message with his secretary.

Kendall did not return to his office until late in the day. He saw the message from Hench, but in his own absentminded way forgot about it. Kendall was likely preoccupied that Friday afternoon; he was leaving the lab early to spend the weekend at his cottage on Lake Zumbro. He departed without returning the call.

At his cabin later that evening, well after the sun had set, Kendall suddenly remembered the message from Hench. His initial thought was to wait until after the weekend and return the call when he got back to Rochester. But for reasons he was never able to fully explain, Kendall felt increasingly guilty as the night progressed. He later described sensing "urgency" in Hench's phone message. Unable to get it out of his mind, he decided to return the call that night. There was no phone in his cabin, so Hench headed down the road to a farmhouse two miles away where he knew he could find one. After knocking on the door, the farmer invited him in— delighted to accommodate someone from the famous Mayo Clinic. Kendall found himself standing in front of an old-fashioned crank phone hanging on the wall. He had to stand uncomfortably during the ensuing conversation in order to be able to speak directly into the mouthpiece mounted on the unit.

Hench was not known for the brevity of his conversations, and this one was no exception. For nearly an hour the rheumatologist talked incessantly. He recapped the entire story involving Mrs. G. and the failed lactophenin treatment. He explained her refusal to leave the hospital. Hench acknowledged that he was sympathetic to her situation and desperate to offer her something that might be helpful. Would Kendall be willing to supply some compound E—as they had discussed almost a decade ago—to try as an experimental treatment for rheumatoid disease?

Kendall was also sympathetic, but he told Hench that he could not provide him with any compound E unless it was approved by Merck— the material he possessed had been manufactured by Merck and wasn't his to give away. He was, however, strongly supportive of the idea and would help Hench obtain the necessary permission and, if possible, an additional supply of the compound. More important, according to Kendall's recollections, "the conversation was concluded by my promise to provide sufficient compound E to treat one patient" if Merck did not.[4] In other words, if all else failed, Kendall would tap into his supply of bovine adrenal glands and extract the necessary compound E to treat Hench's patient.

The phone call cost $1.14. Kendall reimbursed the farmer from his own pocket before departing on the long, late-night walk home. As he meandered back, Kendall wrestled with second thoughts about the commitment he had just made. Although they had discussed it in the past, there was at most an unsubstantiated, fragile rationale for using compound E in this situation. A trial of compound E in this setting clearly involved more imagination than science. And "imagination," Kendall once wrote, "is a phenomenon that is the basis of creative genius. It is an essential ingredient of progress, and yet it can lead to the generation of hope and expectation that have no substance in fact. Experience can develop the power to control one's imagination."[5] Kendall wondered to himself if this wasn't a good time to let their vast experience put the brakes on his—and Hench's—imagination.

Returning to work on Monday, Hench and Kendall, along with Slocumb and Polley, began to devise their strategy for obtaining the necessary supply of compound E. Kendall contacted Merck & Company, explained the situation, and asked for some of the drug. The Merck answer was a waffling "maybe." The company requested that Hench send a letter outlining his rationale for testing the drug in this situation.

The team knew they had one chance. They spent several days and wrote several drafts before sending a letter on September 4, 1948, to Dr. Augustus Gibson, assistant medical director at Merck & Company. The letter pointed out that "we have an ideal patient here now to use as a first case. She is a rather severe rheumatoid arthritic whom we have had under observation and upon whom we have made many clinical tests."[6] Hench kept the medical rationale as simple as possible, largely because the logic was so convoluted that it could not withstand careful scrutiny. "Because of the ameliorating effect of pregnancy, some have incriminated disturbances of hormones; some have incriminated the pituitary, etc. Because of the ameliorating effect of jaundice, we and others have studied carefully the liver. Clinically and histologically from time to time, the adrenals have been incriminated, chiefly because most rheumatoids tend to have a low blood pressure, are tired and weak and many of them develop slight brownish pigmented spots on the skin." Knowing that the company could and would commit, at most, the smallest possible amount of compound E that might be effective, Hench also took care to explain that it shouldn't take long to see effects if compound E turned out to be effective. "We know that there is a potentially provocable mechanism which is activated by pregnancy and jaundice very rapidly. Therefore, if any adrenal compound is of real significance in rheumatoid arthritis, we would expect to see some results within a very few days." With a lick and a first-class stamp, the letter went into the U.S. mail.

The answer came quickly. In the words of Dr. Polley, "The three of us agonized over the contents of the letter for about a week, but obviously it proved to be sufficiently convincing."[7] Merck agreed to Hench's request. Merck was going to ship 5 grams of compound E, an amount worth, at that time, more than $1,000.

For Sherlock Holmes, an "investigation," whether it involved a crime, a mystery, or a scientific problem, was merely a "game." For Hench and Kendall, the game was about to be played.

A package containing 5 grams of compound E arrived from Merck as expected. But before the first dose could be drawn up, the experiment almost came to a quick end. Years later Hench gave an interview for the *Saturday Evening Post* in which he described how he and his team "almost went into shock" when the valuable bottle of compound E was unceremoniously dropped on the hard marble floor of the Plummer Building. It was a miracle that the glass didn't break, turning the potential wonder drug into junk for the janitor.

Now they had a small supply of compound E, but several questions still needed answers before testing could begin. What chemical form of the compound should be used? How should it be administered? How much should be given? In each case Kendall and Hench had to guess, and each guess had critical implications. Hench surely knew of Sherlock Holmes's opinion about guessing. In *The Sign of Four*, Holmes admits that "I never guess. It is a shocking habit—destructive to the logical faculty."[8] But in this case there was no alternative—they would have to make a "best guess."

What chemical form of compound E should they use? It was up to the chemists to decide the answer to this question. Kendall knew that the greatest solubility would likely occur with the "free form" of the drug—in other words, the form of the drug left after the acetate group was removed from their specimen of compound E. A process to do this had been devised by Dr. Vern Mattox of the Mayo Clinic; it was effective, but unfortunately produced only a 75 percent yield. Four grams of their valuable sample would therefore wind up as only 3 grams of free compound E after the conversion. At $200 a gram (in 1948 dollars) this represented a painful loss. Nonetheless, a decision was made to proceed in this way. A solution of the drug was prepared by dissolving 1 gram of acetate-free compound E in 50 cubic centimeters of saline.

How should the drug be administered? Intravenously? By intramuscular injection? By mouth? The answers depended upon chemical characteristics (such as solubility) and physiological factors (like intestinal absorption); for this reason the chemists and the physicians collaborated on the decision. After considering all factors, they decided the most reliable method was by intramuscular injection.

How much should be given? This was the biggest question of all. Their supply was extremely limited, so it behooved them to give as little as possible in order to make it last as long as possible. On the other hand, they wanted to be sure that they gave enough—if the dose was too small and no effect was seen, they would be unlikely to get another chance of using more drug. If the injection was effective, they reasoned, additional studies to determine the utility of smaller doses could be conducted.

Hench was busy, so he delegated this decision to his colleague Dr. Slocumb. Knowing that Sprague and his endocrinology colleagues had been giving compound E to patients with Addison's disease and diabetes, Slocumb sought their advice. Unfortunately, every Mayo endocrinologist was out of town at the time, and he needed an answer now. He discussed the situation with other clinical colleagues, but no one had any strong opinion. In desperation, Slocumb sought advice from Kendall—the ultimate nonclinician.

Kendall pondered the question and offered his opinion. Based on earlier work using adrenal extracts, Kendall believed that 50 milligrams of compound E a day was the largest dose that had ever been used in humans. He suggested they double this to 50 milligrams twice a day, thus ensuring that the dose would likely be large enough to produce an effect.[9] Kendall was keenly aware of the importance of this decision. "If 50 milligrams free form (twice a day) intramuscularly doesn't work, then E doesn't work."

Furthermore, Kendall was sure that "it was not probable that any other method of administration would give results that were more encouraging."[10]

Kendall, who tended to be overly optimistic and rarely depressed—even when optimism seemed foolish and depression was appropriate—couldn't help but feel jitters as the time to begin the trial approached.

> The clinical trial of compound E was similar to that of compound A, except in one respect. In the case of compound A it had been possible to justify failure of activity by the thought that, after all, this was only compound A and we could still look forward to the clinical trial of compound E. But in September 1948 we had in hand that elusive hormone, compound E. We had reached the end of the road. The answer had to be yes or no.

The outcome of this experiment, Kendall knew, would likely be the defining event of his life. "It is difficult to contemplate in a detached and impersonal manner the remaining days of one's scientific career . . . after 18 years of interrupted effort, such a time had arrived in my life.[11]

After almost two decades of scientific commitment to this project, the next week would reveal whether Kendall would be a hero. Or a bum.

On the afternoon of Tuesday, September 21, 1948, Dr. Polley was walking out of the Plummer Building when he overheard three Mayo endocrinologists talking among themselves. He remembers one of them saying, "They are giving compound E to a patient with rheumatoid arthritis today!" He described the tone of voice as one "which we have never been able to duplicate, but it was clearly indicative of a what-do-you-suppose-they'll-think-of-next attitude."[12]

Polley was mildly offended, but he held his tongue. He understood that compound E was outrageously expensive, and that reasonable people were going to be annoyed if they thought it was being used "in any irrational or irresponsible sort of way." He knew he had to expect—and understand—that others were going to feel negatively about this experiment.

Dr. Polley looked down at his watch and noted that the endocrinologists' remarks were being made at the exact moment when the first dose of compound E was scheduled to be injected into Mrs. G.

CHAPTER 20

The Amazing Mrs. G.

We also have our diplomatic secrets.

—SHERLOCK HOLMES,
THE ADVENTURE OF THE SECOND STAIN

MRS. G. RECEIVED HER FIRST INJECTION OF COMPOUND E IN THE
early evening on September 21, 1948. Dr. Slocumb provided the 50 mil-
ligram injection; he was in charge of Dr. Hench's hospital service, and would
be overseeing the day-to-day care of Mrs. G. and the other patients hospitalized with rheumatological disorders.[1]

Where was Hench? As usual, he was preparing for a trip. He'd be leaving
in seven days for London, where he would give the prestigious Heberden
Lecture.[2] Coincidentally, Hench's upcoming talk was titled "The Potential
Reversibility of Rheumatoid Arthritis." As the date of departure approached,
Hench shifted into hyperkinetic mode; because of his speech impediment,
the text and slides had to be perfect. The rheumatologist was both busy and
distracted.

Mrs. G.'s first dose of compound E was, as noted earlier, the culmination
of eighteen years of work for Dr. Kendall and possibly the defining moment
of his career. The gravity for Hench was not quite as serious. Although he
was interested in compound E, in many ways it was just another potential
therapy for rheumatoid arthritis, no different than dozens of others he had
investigated earlier. The sad truth—compound E was something to try on
a problematic patient for whom there was nothing else available. Hench
wanted it to work, but deep in his heart he knew that it, like all the other
therapies he had tried, probably would not. And with the demands of his

upcoming trip pestering him incessantly, there was no way he could make Mrs. G.'s problems his main focus right now. Mrs. G. surely felt differently about her situation. At that time there were an estimated three million patients in the United States with rheumatoid arthritis, but she was certainly one of the most severely afflicted. Young, pretty, and at first glance healthy appearing, the petite woman was completely debilitated by her disease. "The patient could hardly get out of bed; once she tried to walk, it was too painful, so she remained at rest."[3] For Mrs. G., day-to-day living was little more than an intolerable, subhuman existence.

September 22 came and went. Mrs. G. noted no effect from her injections. Dr. Slocumb did not even bother to make a note on her chart.

On September 23 the patient awoke feeling a little better. Amazingly, over the course of the day she seemed to improve even more. Mrs. G. reported feeling less stiff, although on physical examination her joints were still inflamed. Dr. Slocumb wondered if this was merely a placebo effect? That evening he discussed the situation with Hench, insisting that "you've got to see Mrs. G. tomorrow before you leave town and see if she's still the same."[4] When Hench balked at the suggestion, Slocumb, not usually one to challenge his boss, dug his heels in and demanded that Hench visit Saint Marys Hospital and lay eyes on the woman.

The following morning, September 24, it was apparent that something quite extraordinary was happening at the hospital. According to Kendall, "we found her exercising, raising her hands over her head . . . previously impossible. She visited several patients to demonstrate her changed condition. Painful stiffness was gone." As the patient herself noted, "My muscles feel stronger, and my appetite is very good."[5]

As hard as it is to imagine now, Philip Hench was so preoccupied with his impending trip to England that he still found it difficult to make the half-mile trip from his office to the hospital in order to see the ongoing transformation for himself. Perhaps he thought his colleagues were exaggerating Mrs. G.'s improvement. Or maybe he was just so focused on his travel preparations that he'd managed to compartmentalize his problematic patient in some far-off mental place where she could not distract him. It was with very great reluctance that Hench finally agreed to visit Mrs. G. at 7:30 P.M. on the night of September 24.

Dr. Hench was in a hurry. The patient was in room 127 at Saint Marys Hospital, located in a bed next to the window of a double room on the first medical unit. The room was directly off a stairwell, and Hench was anxious to get the visit over with. "Dr. Hench surprisingly came up the back stairs of the hospital two steps at a time, whereas he usually didn't even walk up stairs one at a time, if there was an elevator available."[6] As Polley notes, "Phil . . . was about six feet four and weighed about 220 pounds and he was a very hyperkinetic fellow, and we used to say he moved a lot of air."[7]

Reaching the appropriate floor, Hench turned into the room and began waving his arms with seemingly comical vigor. But this time the usually humorous doctor wasn't trying to amuse anyone. Hench was stressed, hurried, and in a rare display of ordinariness, somewhat rude. "You're ruining my evening!" he announced to the patient and any staff within earshot.[8] And this time Hench, a notorious joker, was not kidding.

But over the ensuing minutes, he began to appreciate the magnitude of Mrs. G.'s clinical evolution. She was a woman whose physical limitations were well known to Hench; he had dealt with them for many weeks, and they'd been highly refractory to every form of treatment he had tried. When he last saw her over a week ago, she was completely debilitated. The woman in front of him now was an entirely different person. She stood, walked, and lifted her arms with ease. Hench was flabbergasted. As the significance of this metamorphosis sank in, Hench began to feel a strange tinge of remorse. He was about to leave for London. He would not be able to observe or study her clinical course. Mrs. G.'s so-called cure might be the biggest thing ever to hit the field of rheumatology, and he was about to miss it. As Polley mused, "After visiting her for a while, however, it became apparent that she was unwittingly ruining a lot more than his evening."[9]

Within one week of the initiation of therapy, Mrs. G. was virtually pain free. Kendall reported that "after six days she had lost almost all her stiffness; articular tenderness and pain on motion were markedly reduced. The next afternoon (September 28) she shopped for three hours downtown, feeling tired thereafter but not sore or stiff."[10] She noted a sense of well-being: "I have never felt better in my life." Hench, still reeling from the "turmoil and emotion" in his mind, maintained the good sense to share the credit with Kendall, who had been following Mrs. G.'s progress through reports provided by the hospital staff. Hench invited his colleague over to Saint Marys and introduced him to the patient. Kendall, a chemist and not a physician, rarely had an opportunity to meet the patients he helped. He was genuinely touched when Mrs. G. rose from the bed and said, "Let me shake your hand."[11] Kendall "never made any big deal out of it,"[12] but those who knew him realized just how much he appreciated this simple encounter.

Hench and his team were now locked in nonstop discussions about Mrs. G. As he prepared to embark for Europe, Hench raised an issue that seemed inordinately strange to his partners. Secrecy. Hench was absolutely insistent that the team keep their findings confidential and avoid premature publicity. To the others, this was "a paradox"—after all, compound E was "a matter of record," so why was there a need to keep this wonderful discovery under wraps? But Hench was adamant—and he was the boss. This was going to be their little diplomatic secret, and while he was away for the next ten weeks there was to be no public or private disclosure of their work. He would correspond with them frequently to keep up on progress. But—and now Hench was beginning to sound almost paranoid—he wanted the correspondence conducted in code.

The team agreed to go along with the ruse. They decided that compound E needed a secret code name.[13] Polley suggested they switch "E" to a different letter—perhaps one common to all their names? Unfortunately, the only letter meeting this requirement was "E." Slocumb offered "H" as an option; it was common to three of the four names, and also invoked images of "Hench, hormones, and hallelujah." The team agreed that they would refer to the new substance as "H." They would contact one another in writing only. Mail and telegrams were fine, but the phone, still a suspect instrument of communication, would be avoided.

CHAPTER 21

A Promising Start

American slang is very expressive sometimes.

—SHERLOCK HOLMES,
THE ADVENTURE OF THE NOBLE BACHELOR

DR. AND MRS. HENCH DEPARTED FOR THEIR TEN-WEEK TRIP TO ENG-
land on September 28, 1948, just seven days after Mrs. G.'s first dose of
compound E. Although Hench's trip had been set in stone for months, the
events of the previous week mandated a slight change of plans. He made a
short side trip to New York City en route to London, and there he met with
two important corporate people: Dr. James Carlisle and Dr. Augustus Gibson.[1] Dr. Carlisle was the medical director of Merck, and now had primary
responsibility for the compound E project and the huge ($14 million) invest-
ment his company had made in it so far.[2] Dr. Gibson, Merck's laboratories
director, had supplied Hench with the original 5 grams of compound E.

Hench's visit was more than a chance to thank his benefactors directly.
Without revealing too much information, Hench wanted to foster continued support for the project. Mrs. G.'s stunning response was encouraging,
but Hench—in sharp contrast to Kendall—knew the tremendous downside
to a premature (or, God forbid, incorrect) announcement of success. A lot
more work needed to be done before any claims were made. It was largely
up to Hench to convince Merck that it should support the work financially
until it could be heralded in public.

The details of his meeting with Merck are unknown, but Hench, always
persuasive, must have worn down any opposition to his plans. A deal was
struck in which Merck agreed to supply additional compound E for up to
four more patients. But there was a catch. Dr. Carlisle himself wanted to

travel to Rochester and see the drug's effects on the next patient or two. Hench agreed, passed the news back to his colleagues in Rochester, and headed across the Atlantic to give the Heberden Lecture in London.

The first patient to receive compound E had been a woman. With the blessing of Merck, Polley and Slocumb decided that the next two patients should be men. Two additional patients were identified and admitted to Saint Marys Hospital, where they shared a double room. They were an odd couple. The older gentleman, Mr. W., was, at least to the extent that any patient can be considered typical, a fairly typical patient with rheumatoid arthritis. Polite and cooperative, he was extremely debilitated by his disease; he'd be quietly appreciative if this new therapy turned out to provide any benefit at all. In contrast, his younger roommate, Mr. M., seemed to be a difficult, manipulative patient—far more crotchety and frustrating than the older man.

Both men were essentially bedridden when therapy with compound E was initiated. The administration and dosages were the same as those used on Mrs. G., and within a few days there was a similar—and striking—improvement in both men. What differed markedly between the two was their attitude about this stunning change. Mr. W. readily admitted that he was feeling much better with the treatment. But Mr. M., even though he was now frequently jumping out of bed to assist his more severely afflicted elderly roommate, stubbornly refused to concede that he'd had any improvement at all. All the doctors and nurses watching his progress could easily see the changes that were taking place, and they were amazed by the apparent discrepancy between their observations and Mr. M.'s perception. This scene went on for a few more days—Mr. M. prancing about the room while adamantly denying that he was the slightest bit improved. Charles Slocumb finally decided to get to the bottom of this odd behavior. Exactly one week after starting treatment, Slocumb asked Mr. M. during morning rounds whether he was feeling any better yet. When told "no," he suggested that "maybe we shouldn't waste injections if they aren't helping?"

Mr. M. collapsed suddenly into his chair and a panicked expression crept over his face. "No, no, no!" the young man protested. There was more fear than anger in his voice.

Dr. Slocumb sat down next to him and gently began to probe the situation. Mr. M. reluctantly explained that he had "been afraid to accept the improvement as being genuine until it had been sustained." On too many occasions in the past, he claimed, he had thought that a particular new treatment had bestowed a "therapeutic" response, only to find that the effect was transient, a placebo phenomenon, or perhaps even a total figment of his imagination. And although he was terrified to admit it, compound E seemed to be working far better than anything else he had ever tried. Mr. M. was scared that if he acknowledged this benefit, the drug would, like previous therapies, stop working.

Mr. M.'s situation was something that Slocumb and his colleagues understood very well. "All of us were well aware of what Hench frequently used to refer to as the 'inevitable 65 percent'—that large percentage of favorable responses that seem to occur, initially, with most therapies for rheumatoid arthritis." Slocumb reassured Mr. M. that the drug was truly working. The young man, in turn, agreed to work "honestly" with his physicians for the duration of the trial.

Luckily, Dr. James M. Carlisle of Merck & Company never heard Mr. M.'s early claims of "ineffectiveness"—he'd have likely reacted by gleefully crushing Mayo's fragile compound E research program.

Carlisle arrived in Rochester on a cool fall morning following an overnight train ride from Chicago. He made an immediate impression on Hench's colleagues. Polley, in published remarks, simply stated that "his icy detachment and frigid skepticism rivaled the temperature at that time."[3] Polley's unpublished comments are blunter. "Dr. Carlisle's assumed affect (a nice, cold affect at that)" was striking. "He arrived acting like the world's most skeptical skeptic." Apparently Polley's observations were in line with those of his colleagues. "Dr. Slocumb remembers the day of his arrival as 'a very bad morning.'"

It was assumed that Carlisle's attitude stemmed from the heated "discussions" that had taken place weeks earlier regarding the amount of compound E that Merck would agree to supply. Slocumb was unwilling to initiate the second and third patients until he had a guarantee from Carlisle that adequate compound E would be provided by Merck so that they could taper the dosages in Mrs. G. and the two new patients gradually if it became necessary to discontinue the drug. Since it was completely unknown at this point whether—or for how long—a tapering dose would be necessary, this amounted to an open-ended commitment to supply the drug. Carlisle said "no." It was only when Slocumb turned to Kendall, who agreed to manufacture a supply of compound E for these patients if Merck refused, that Carlisle finally capitulated.

For the Mayo investigators, dealing with Carlisle was only slightly better than having a red-hot knitting needle poked into your eye. It seemed that he enjoyed being difficult. The uncomfortable relationship between Carlisle and the Mayo team culminated one evening in the room of Mr. M. Carlisle knew he wasn't supposed to be in Mr. M.'s room; he had "sneaked" into Saint Marys Hospital without Polley's or Slocumb's knowledge and was surreptitiously visiting with the patient. The conversation focused entirely on the effects of compound E. Claiming that he was merely serving as the devil's advocate, Carlisle attacked compound E with unconcealed bitterness. In a manner as deprecatory as possible, he belittled "the anti-rheumatic effects as bitingly as he could." Carlisle argued that the compound E trial was little more than some type of fraud being perpetrated on Merck, and he was happy to take Mr. M. to task for his role in the sordid affair. Mr. M. repeatedly denied these accusations and told Carlisle that the drug was working as well as everyone claimed.

Mr. M.'s cantankerous nature emerged in this setting, and when he couldn't—or wouldn't—tolerate it any longer, he spoke out. "Doc, you may be a good doc where you come from, but as far as I'm concerned, you're full of . . ."[4] Dr. Polley tactfully noted that the word Mr. M. used was "a four-letter word for lower bowel contents now classified by the U.S. Supreme Court as 'indecent.'"

The blunt response floored the otherwise unflappable Dr. Carlisle. Like Saul walking the road to Damascus, Carlisle underwent a conversion on the spot. He was instantly turned into an advocate for compound E. According to Polley, "We have often speculated that this forthright retort might have had a favorable influence on the pharmaceutical company's decision to proceed with the studies."

In the days that followed it became clear that Carlisle was, at his core, a decent person. His aggressive, irritating behavior had merely been an act to ensure his absolute objectivity in the assessment of compound E. After spending $14 million of Merck's money on this project, it was up to him to turn off the pharmaceutical life-support system if the drug wasn't working. He was now convinced that it did work. Carlisle agreed to continue his support for the project and from that point forward was "agreeable, reasonable, and cooperative." Amazingly, "we all quickly acquired the greatest and enduring admiration for him personally and professionally." Carlisle and the Mayo team became friends, and would remain so until death finally separated them.

Dr. Carlisle had started out as an obstacle to Mayo's compound E team, and he wasn't the only Merck official to make waves for the project. One morning in 1948 Dr. Hans Molitar, one of Merck's research directors, arrived unannounced at Saint Marys "obviously with chips on both shoulders." Polley knew very little about the reason for this surprise visit.

"We had heard some rumblings to the effect that he and the Medical Director [Carlisle] had what might be called 'a conflict of interest' and that's about all we really knew about it." Molitar, who was "prancing around the floor in a very restless way," found Dr. Slocumb making rounds and rudely interrupted him. "What's going on here?" Molitar demanded to know. Before Slocumb could answer, questions began to fly like Tommy gun fire. "How does compound E work in rheumatoid arthritis? What causes rheumatoid arthritis?" Slocumb, his patience tested, responded "I don't know" to each question. He offered to show Molitar some patients with the disease. The director refused. Slocumb offered to sit down with the director and discuss the questions in depth. The man from Merck was apparently not interested. He "turned on his heels and left in a huff and that was the only time we ever saw him in Rochester."

Despite his conversion, Dr. Carlisle, now considered to be a "liaison, and . . . friend," was still not always the easiest man to deal with. He wasn't afraid to agitate the Mayo team when it served a good purpose. Polley reports that during one of his visits to Rochester, Carlisle spent almost a week studying the charts of patients receiving compound E. After ignoring the team members for several days, he confronted Kendall with a bold accusation: Carlisle's calculations suggested that the Mayo team had used several grams more compound E than should have been possible based on the amount that Merck & Company had supplied to them.[5] Kendall simply laughed. It turned out that he and Dr. Slocumb had made arrangements to flush every syringe, needle, and bottle in an effort to recover all possible compound E residue. When this was added into the equation, it easily accounted for the discrepancy. Carlisle, now much more at ease, remarked, "It's the reverse of what I usually have to put up with." Trust was continuing to slowly develop between the Mayo researchers and the businessmen of Merck.

Between November and December 1948 the first three patients continued their treatment with compound E while two new patients initiated treatment. Patient number four was a Mayo Clinic staff neurologist. He was thought to be a perfect candidate; his disease had been extremely well studied (owing to his affiliation with the clinic), and he, unlike many of the lay patients, could communicate with his caregivers using modern medical terminology.

The rheumatoid-stricken neurologist received compound E for twenty-four consecutive days, during which he experienced a "comfortable wakefulness" no less astonishing than the relief seen in the first three patients. He was able to complete three medical manuscripts that had remained unfinished for the past two years. He moved about the halls and climbed stairs with relative ease. Sadly, when his compound E was eventually discontinued, he experienced "depression and pain."

The impending treatment of patient number five posed a dilemma. By now, the Mayo board of governors was aware of the initial success with compound E, although the rheumatologists were desperate to honor Hench's request to keep the studies secret until he returned from England. The board members took it upon themselves to ensure that the testing on patient number five would be "as convincing as possible." A five-person committee was appointed to help design a protocol for definitive testing.[6] They decided on a "double-blinded placebo-controlled" study.

Kendall was asked to make an injectable placebo that resembled compound E, which he prepared using a cholesterol solution. Dr. Sprague was the only person who would know which solution contained compound E and which was placebo. Three clinicians, none of whom knew which drug was being administered at any given time, were instructed to examine the patient daily, and to independently note their findings. In Hench's absence, Drs. Manfred W. Comfort (who had been designated as liaison by the Mayo board of governors), Dr. Sprague, and Dr. H. L. Smith were placed in charge of the trial. Patient number five was admitted to the new Saint Marys Hospital Metabolic Study Unit (a new facility specifically designed for investigational studies of this type), and the first dose of compound E was administered on November 28.[7]

Patient five owned a trucking transportation business. He had recently come to the clinic because of severe musculoskeletal pain, and had just received a diagnosis of rheumatoid arthritis from his Mayo doctors. Treatment with aspirin helped his symptoms a little, but not enough to allow him to return to work. He was miserable, and his disease was relentless.

For the first eighteen days of the trial the patient was given aspirin only. No injections of either compound E or placebo were administered. The patient's clinical status did not improve. Between December 17 and 23 he received five injections of the cholesterol placebo. Based on the first four patients, five days of injection should have produced a significant improvement in his condition. It didn't. The patient's clinical status still did not change.

On December 24 he received his first injection of compound E. The second dose was given Christmas morning. That afternoon, a phone call was made to Kendall while he was out driving with his wife. When he returned home, he glanced over the hand-written message that had been left for him. His mind drifted back to the Christmas day in 1914 when he had first crystallized thyroxine in his laboratory. The message in front of him read simply, "Miracle number five has happened at Saint Marys Hospital."[8]

A miracle indeed. Patient five was essentially pain free after only two injections. He climbed out of bed and walked the corridor. He climbed stairs. Most impressively, he expressed a desire to get back to work as soon as possible.

All five members of the board-appointed committee were asked for their opinion. Is compound E effective in the treatment of rheumatoid arthritis? A consensus was reached. It is. Even the most skeptical member of the five-man committee admitted, "I capitulate."[9]

CHAPTER 22

The Bad and the Ugly

There is but one step from the grotesque to the horrible.

—SHERLOCK HOLMES, *THE ADVENTURE OF WISTERIA LODGE*

THE APPARENT SUCCESS OF CORTISONE IMMEDIATELY CREATED PROB-
lems, and honoring Hench's instructions to keep the trial a secret from the
public—and his professional colleagues—was among the most pressing of
these. A simple error in logistics was threatening to unveil their clandes-
tine activities. What error? Hospital room assignment. Mrs. G. had been
given the wrong hospital accommodations for an experiment of this type.
The team had not anticipated that the effect of compound E would be so
dramatic. Naively, they had put her in with a roommate. And now this
roommate had witnessed everything.

Mrs. G.'s roommate had been referred to Mayo by a well-known Chicago rheumatologist who was a close personal friend of Polley, Slocumb, and
Hench. She'd observed Mrs. G.'s impressive transformation, and was privy
to all of the discussions that took place at the bedside, including the initial
informed consent communication and the subsequent follow-up reports.
She knew exactly what was going on. Something called compound E was
miraculously curing Mrs. G. Unable to receive compound E herself (due to
the limited supply), she was going to be sent back to her referring rheumatologist shortly.

Nobody wanted the premature announcement of compound E's amazing abilities, especially if the compound wouldn't be available to the public
in the foreseeable future. It seemed criminal to create a demand for something that didn't exist. What to do?

Slocumb and Polley opted for the direct approach. They sat down with the roommate and explained the situation as clearly and honestly as possible. Then they asked her for a specific favor—that she "say nothing to anyone at anytime about what she had seen or heard—and especially not to discuss the matter with her home physician." In retrospect, it was an extraordinary and potentially dangerous request—they were basically asking the patient to lie to her physician at home. But this nonetheless seemed like the best approach to take. "We could only hope that we had been effective at emphasizing the importance of her cooperation."[1]

A second near breach of secrecy involved another patient treated with compound E. The patient, also a young woman, was a graduate student in biochemistry; her training made her familiar with the nuances of steroid metabolism. Shortly after receiving her first injection, she was asked to provide a twenty-four-hour urine collection (a common procedure on the Metabolic Study Unit), and she immediately recognized from the labels on the jug that the laboratory would be analyzing her specimen for steroids. She mentioned to another patient on the unit that her injections "must have something to do with the adrenal gland." The second patient was a registered nurse at the clinic. She knew a little background on the secret project, and immediately cautioned the young biochemistry student, "You'd better be quiet about it or we'll all be in trouble!"

But keeping the compound E trial under wraps was not the biggest problem the Mayo team now faced. Something completely unexpected—and grotesque—suddenly threatened to undermine everything they were trying to accomplish.

The crisis? Mrs. G. was going crazy.

Looking back on the situation twenty-five years later, Polley and Slocumb's account of Mrs. G.'s side effects seems sanitized. "Our first patient's early course on compound E provided us with the first evidences of withdrawal effects, of the initial euphoric responses . . . and of the need to taper reductions of dosage gradually."[2] It sounds fairly benign. But their unpublished notes paint a much different story.

The team had planned to treat Mrs. G. with 100 milligrams of compound E daily for six months before reassessing her rheumatoid status. As previously described, the first week or two were incredibly successful; her symptoms nearly resolved. But as they did, there was a subtle yet definite change in her mood.

Along with her increased physical capacity came an increase in her psychosomatic state. Her reflexes were set on hair-trigger, and her mentation quickened. Over the next couple of weeks Mrs. G.'s optimism and enthusiasm evolved into euphoria, and then spun rapidly into hypomania. At this point physical deterioration began to set in. "She got gray, cushingoid, coarse and depressed and psychotic." Her body changed. She developed a bloated, "moon" face, and streaklike lesions on her body (called stria). In a matter of days her mental status worsened; she became depressed and needed to be transferred to the closed psychiatric ward for urgent management. The frightening change from a person crippled with arthritis to an effervescent gadabout to a psychotic, bloated, fragile mental patient took no more than a month.

Mrs. G.'s mother was a regular visitor to the hospital during the compound E trial. What she saw happening to her daughter now discouraged her enormously. She "became very estranged from us [Polley and Slocumb] and claimed that we ruined her daughter's life." The mother, according to Mrs. G.'s doctors, had literally "turned against Mayo," and her mother forbade any further contact with the clinic's doctors.

Mrs. G.'s compound E dosage was slowly adjusted, her condition improved slightly, and she was eventually able to leave Saint Marys Hospital. Despite her marginal mental status, she began keeping a diary after her discharge from the hospital. The fate of this diary remains unknown. Mrs. G. returned to Kokomo, Indiana, where she followed up with her local doctor in Indianapolis. He was amazed by what he saw—she now appeared to have a condition similar to systemic lupus erythematosus.[3] "Apparently, this further disenchanted her mother who thought or allowed herself to think that we had treated her for rheumatoid arthritis and that wasn't what she had." It surprised no one when Mrs. G. discontinued using compound E and refused to ever take it again.

Over the ensuing years Mrs. G.'s arthritic condition worsened, leaving her once again completely debilitated. In 1954, during a flair of her disease, she was hospitalized at Methodist Hospital in Indianapolis. Her doctors wanted to treat her with compound E, but she adamantly refused. She had been told years earlier, when it was suggested that she actually had lupus rather than rheumatoid arthritis, that adrenocorticotropic hormone (ACTH) (a pituitary hormone) would be a better treatment for her. Her Indiana doctors compromised and provided her with the hormone.

Disaster ensued. The ACTH stimulated her adrenal glands and caused them to produce a DOCA-like substance that had minimal effect on her inflammatory disease but caused her to retain salt and water. Her lungs filled up with fluid. Mrs. G. died of acute pulmonary edema in December 1954, drowning in her own steroid-induced buildup of fluid. Her mother, still bitter about her daughter's outcome, refused an autopsy. With no autopsy or diary to provide additional "data," Mrs. G.'s final years were reduced to a scientific vacuum.

It was apparent that compound E wasn't a cure for rheumatoid arthritis. When the drug was stopped, symptoms inevitably returned—and every patient eventually stopped taking the drug, either because the supply ran out or side effects developed. Efforts were made to determine whether smaller doses would still produce the desired effect, but when less of the drug was given, the clinical response was often unpredictable. High doses and the significant side effects they produced remained the rule rather than the exception.

In short, Mrs. G.'s clinical course was by no means unusual. Consider the frighteningly similar story of Janis Brown.[4]

In February 1949 Janis Brown was a sixteen-year-old high school student in northeastern Iowa. Following an upper respiratory infection, she developed diffuse, debilitating muscle and joint pain. Her father brought her to Saint Marys Hospital, where she was diagnosed as having acute rheumatoid arthritis. Mrs. G. had become the first person to receive compound E for this condition just a few months earlier. Janis Brown, however, was not treated with compound E due to its limited availability. Instead, she received some of the other novel therapies that were being tried at the time.

The first was 17-hydroxy progesterone (a steroid similar to progesterone that had recently become available in intramuscular form). After two weeks of daily administration, the drug had shown virtually no effect. She was subsequently treated with ACTH, the same commercially available pituitary extract that had led to Mrs. G.'s demise. Janis Brown fared better; her arthritis improved within twenty-four hours, but the drug had to be discontinued because she developed a rash and hives.

On September 15, 1949, almost seven months to the day after the onset of her initial symptoms, Janis Brown started therapy with compound E. It was initially given at a whopping dose of 150 milligrams twice daily, and then reduced to 100 milligrams a day. By now, Polley, Slocumb, and the others knew what to expect, and Brown did not disappoint them. Within one day she was able to raise herself from her bed without assistance. She described the effect as "miraculous"—her pain had been reduced by at least 50 percent. Four days later Polley sent a telegram to Randall Sprague, who was on a trip to New York City; it read, "[The patient is] responding quite satisfactorily."

But with a course frighteningly similar to that of Mrs. G.'s, the complications of compound E began to develop. By October 16 it was noted that Janis Brown had developed fullness in her face—a "moon" face—which remained evident until the mid-1950s. Her doctors cut the dose of compound E down to 80 milligrams a day, but despite this reduction her skin began to develop stretch marks. Observing this dramatic cutaneous change, her physicians decided to abruptly discontinue compound E. Bad move. "Ms. Brown . . . crashed . . . and wondered if she wasn't worse than before taking (compound E)." Other therapies, including aspirin, physical therapy, and morphine, were reinstituted, but they did little to alleviate her rheumatoid symptoms. By the time Brown left Saint Marys Hospital to return home for Christmas, "she was 'really hormonally out of whack and depressed' and so traumatized by her experience that she could not bear to return to Mayo Clinic again."

Years later, images of the experimental therapy still haunted her.

I'd become overly traumatized by my test patient experience. There were nightmare-type memories of unpleasant, frequent tests with large needles—not like the small needles they have today—which were a part of the dreaded bone marrow chest biopsy and almost daily blood withdrawals. As I recall the worst of times, I must also emphasize that though I often wept with pain, fear, and lonesomeness, I tried to fight wallowing and self-pity and staying down.

Janis Brown opted not to return to Saint Marys, and Dr. Polley therefore never saw his patient again. But he never forgot about her. Polley corresponded regularly with Brown's personal physician back in Iowa. At one point he suggested that her physician try using intramuscular gold on her, which at that time was recognized as an acceptable treatment for rheumatoid arthritis. Brown took this only for a short time but stopped it because of side effects.[5] "I just did not want anymore of it as I had a fear of additional side effects that could deteriorate my health even more."

Unlike Mrs. G., Janis Brown was willing to try cortisone again. In early 1951, after it had become available commercially, she went to the University of Iowa and began receiving the steroid on a regular basis. It helped her symptoms, but she nonetheless remained terrified of the drug. "However, the awareness of my severe [compound E] side-effects encouraged me to try very hard to gradually get off of it, as I knew it could ultimately destroy me or cause even more problems later in life. Finally, I won that battle sometime in 1956 and never went back." Her ability to wean off compound E was ultimately attributed in part to estrogen; in an attempt to reduce her daily intake of compound E, her Iowa doctors initiated therapy with the female steroid. She was soon able to reduce her dose of compound E by half without feeling worse. "I think that hormone combination was the beginning of eventually getting off [compound E] altogether."

In Janis Brown's mind, compound E had permanently destroyed her health: "It had already caused additional health problems like chronic fluid retention, severe potassium depletion causing neurological and heart symptoms, and loss of calcium from the bone resulting in my thin bones." Fifty years later physicians would still recognize these symptoms as typical side effects produced by compound E.

Too scared to try other new pharmacotherapies for rheumatoid arthritis like Plaquenil, methotrexate, and so forth, Janis Brown watched her joints progressively deteriorate. She required bilateral hip joint replacement surgery in 1971. In 1978, twenty-eight years after vowing not to return to the Mayo Clinic, she in fact did return there—but only because her orthopedic surgeon from Iowa City had relocated to Rochester. He performed additional joint replacement surgery on her ankles. The new joints worked reasonably well for approximately three years, after which they failed as the bones in her feet began to break down. Multiple surgeries, including "four joint replacements and one fusion," followed. Other surgeries on the fingers and elbows were eventually required, as was a second hip replacement.[6]

Horrible dramas like those involving Mrs. G. and Janis Brown quickly became known to the American public. Media transmission—and manipulation—was inevitable. There's no better example than 1956's *Bigger Than Life,* Nicholas Ray's groundbreaking film on drug abuse.[7] In the movie, James Mason plays Ed Avery, a milquetoast schoolteacher and father who begins to suffer headaches and blackouts caused by "a rare inflammation of the arteries." After his doctors tell him he has only months to live, Avery volunteers for an experimental treatment—a mysterious drug that turns out to be compound E. Avery makes a miraculous recovery and is able to return home to his wife and son. But there's a catch! He must continue taking the experimental drug or his illness will recur. All's well and good until Avery begins abusing the drug, causing wild behavioral changes and catastrophic mood swings. His cure becomes a nightmare, portrayed Hollywood-style to highlight the inevitable horrors of drug abuse (imagine a steroid-inspired version of *Reefer Madness*).[8]

No drug ever raised expectations or dashed hopes more than compound E. Even Kendall, a man who usually seemed unaware of things happening around him, noted that "the effective use of cortisone was retarded for awhile by intemperate and unscientific extremes of exaggerated praise, bitter denunciation, and emotion-laden criticisms, but these reactions were gradually replaced by more reasonable attitudes."

As "reasonable attitudes" began to prevail, compound E emerged as a powerful therapeutic tool for the desperate clinician. But not, unfortunately, before the "bitter denunciation" and "emotion-laden criticisms" described above would inflict serious psychic injury on one of the men most responsible for the drug's success.

CHAPTER 23

Progress and Setbacks

Detection is, or ought to be, an exact science and should be treated in the same cold and unemotional manner.

—SHERLOCK HOLMES, *THE SIGN OF FOUR*

IN MID-DECEMBER 1948 PHILIP HENCH RETURNED TO ROCHESTER from his highly successful lecture series in England. Patient number five, the "make or break" subject upon whom the board of governors wanted to conduct a double-blind placebo-controlled study, had just been admitted to the Metabolic Study Unit at Saint Marys Hospital. This would be Hench's first opportunity to follow a patient through a cycle of treatment with compound E. Dr. Polley, referring to his boss's reluctance to visit Mrs. G. prior to his trip, quipped that this time, "I had lots less trouble getting Dr. Hench to come to Saint Marys Hospital."[1]

But even the amazing effects of compound E were not necessarily irresistible enough to make Colonel Hench deviate from the most regimented aspects of his daily routine. For example, patient number five was given compound E on December 24 and experienced dramatic improvement the next day. The team affectionately referred to him as the "Christmas miracle." Hench surely must have wanted to witness this dramatic transformation firsthand. But he didn't. "Christmas that year was on a Sunday, and Dr. Hench declined to come to the hospital until the following day."[2] In this regard, Hench could not have been more different from Kendall—as he'd demonstrated on many previous occasions, the Mayo chemist certainly thought nothing of pursuing science in the lab on Sunday. Or, for that matter, on Christmas.

After the unequivocal success of patient number five, it seemed reasonable to try another approach on the next patient. According to Kendall, patient number six was treated with ACTH (a pituitary hormone).[3] (Polley remembers this differently; he claims that the first patient to receive ACTH was patient number fourteen—not six—and that the substance was administered in February 1949.)[4] ACTH was known to trigger the release of cortin from the adrenal glands; it was expected to produce effects similar to those of compound E. If so, why use ACTH? Why not simply give the patient compound E? The answer involved an issue of practicality rather than exact science—it was easier to extract this simple proteinlike substance from leftover slaughterhouse pituitaries than it was to extract a complex steroid like compound E from adrenal glands. ACTH, prepared commercially by the Armour Labs of Chicago, was therefore available in much larger quantities than compound E and for that reason was less expensive than adrenal extracts.

But should it be given at all? Not everyone believed that high-dose ACTH would help people with rheumatoid arthritis. Indeed, one theory held that patients with rheumatoid disease produced "an abnormal or altered adrenal product which, if increased by action of ACTH, might *aggravate* the symptoms of rheumatoid arthritis."[5] The detective in Hench believed it was the right time to answer this question in a cold and unemotional manner by scientifically evaluating the clinical effects of pituitary hormone.

According to Kendall, patient number six received 100 milligrams of ACTH injected as a daily dose.[6] By the third day of treatment patient six (or fourteen, depending on whom you believe) improved dramatically, much like those who had received compound E. Based on this favorable response, six of the first twenty-three patients in the compound E series at Mayo actually received ACTH as all or part of their therapy. Although results were favorable in every instance, it quickly became apparent that the actions of ACTH were not exactly the same as those of compound E, and that a different side-effect profile occurred with its use.[7]

In January 1949 Kendall's life became a little easier. Merck & Company began preparing the "free" form of compound E; now the steroid arrived free of the acetate group that was normally attached to it, and Kendall was free of the responsibility for removing it.[8] Within a short time that windfall began to look a lot less impressive. With all the trials now showing a convincing, beneficial effect, the Mayo investigators finally had a chance to assess the original form of compound E—the "nonfree" form with acetate still attached. It turned out to be just as effective on patients with rheumatoid arthritis as the free form. This simple discovery effectively increased the supply of compound E by approximately one-third, since it was no longer necessary to destroy some of it in the process of removing the acetate group.

The next misconception the Mayo investigators cleared up was even more important than the "free" form fiasco. They gave compound E to their patients by mouth rather than by injecting it. To the amazement of all, oral administration worked just as well as injection. The Mayo investigators had initially played it safe by doing things the hard way—preparing the acetate-free form of the drug and administering it via reliable, albeit painful, injections. This had been Kendall's idea, but once again Kendall turned out to be wrong. The much cheaper acetate form, which could be conveniently taken by mouth, was just as potent as the injections of the insoluble acetate-free form of the steroid. These were crucial discoveries, and they greatly simplified the administration of compound E. Years later Kendall reflected that "perhaps a critic would remark that we had overlooked the simplest answer to the problem of administering the cortical hormone and had learned the hard way. However, we had no regrets. Each step was planned and the results were firmly established before we contemplated simplification or change. The hard way, under the circumstances, was the safest and surest way."[9]

After Hench's return, Dr. Polley replaced Dr. Slocumb as the hospital rheumatologist in charge of compound E testing. The next twelve patients were admitted and studied under Polley's care. The investigator may have been different, but the results were the same. As Polley noted, "our exciting and exceedingly time-consuming efforts continued."[10]

The efforts may have continued, but not without some "vexing frustrations," including those brought on by Dr. Hench himself. The jovial, personable rheumatologist was becoming "distressed about what he considered to be interferences and unwarranted restrictions on the progress of the investigations. He seemed increasingly intolerant of criticism of the studies, which he regarded as a personal affront, and was much more interested in the initial anti-rheumatic effects than in the later and quite distressing (. . . side effects)."[11] Unpublished remarks by Dr. Polley describe the first signs of a subtle transformation that may have been affecting Hench at this time.[12] "Phil seemed to begin at once to work out his accumulated frustrations over the months by trying to change everything that the rest of us had been doing while he was gone (to England) and particularly for the sake of reestablishing his captaincy in a position which was never in jeopardy, as far as any of us were concerned or anybody else has ever heard about."[13] Randall Sprague remembers being "verbally harassed for nearly an hour" by Hench because of "an unwarranted intrusion by an outsider in the rheumatologic study."[14] These observations suggest that Hench, at the very least, was feeling tremendous stress over his work with compound E.

If some describe Hench as "intolerant of any criticism of the side effects of (compound E)," others tell a dramatically different story.[15] Dr. Richard Freyberg, the noted New York rheumatologist, knew Hench well and offers an alternative view.[16]

> I can best describe the toxic effects of compound E by telling a story about Dr. Hench. He was a very dedicated physician. Just prior to his presentation at the 7th International Congress, he and I were sitting in his hotel room when he received a call from a woman who had rheumatoid arthritis and was one of the first who had received compound E. She was crying on the phone. Between sobs she told how miserable she was, feeling persecuted, suffering delusions, and unable to sleep. This phone call had a profound affect on Dr. Hench. Before I knew it, he was crying, "What have I done to this patient?" I am convinced that this phone call influenced the tone of his presentation that day. He cautioned the listeners to be alert for toxic effects, calling for careful scrutiny of patients and their complaints after they had received the drug.[17]

The accounts by Polley, Sprague, and Freyberg seem slightly at odds with one another, but they all make the same points. Hench was stressed, and his behavior may have been affected.

One thing that was not changing was Hench's desire to keep the clinical trial secret. This policy was tested during the visit of a Danish exchange professor who was visiting some friends at Johns Hopkins; the man had traveled at his own expense to the Mayo Clinic in order to find out more about rheumatology practice in the Midwest. This type of professional visit was common at the time. Unfortunately, it took place just as the compound E trials were captivating the Mayo rheumatology team's attention. Hench, who was politely hosting the doctor's visit, was desperate to see that his guest had a worthwhile experience, but also wanted to ensure that the visitor found out nothing about the compound E trial. "Phil took him on scenic trips all around Rochester and for miles in every direction and visited with him at length at the Hench home and gave him books to read and arranged conferences with him with other people in the Mayo staff and so forth."[18]

Finally, unable to amuse his guest any longer without jeopardizing the secret experiments that were going on under the Dane's nose, Hench utilized a strategy common today; he pawned off his shadow on an underling. In this case, Dr. Polley pulled the short straw. Hench told Polley to keep his guest entertained, but to make sure the Danish doctor didn't get wind about compound E. Years later Polley would claim that he still "gulps" just thinking about Hench's charge. Like a good subordinate, Polley carried out his boss's wishes. Conducting rounds of the hospital with the visitor and the medical residents, Polley had several patients brought to a secluded examining room where they could be examined and questioned. Eventually the discussion turned to pathology, and Polley was able to persuade his charge to spend the rest of the day in the pathology lab—far from the clinical facility where the patients were cloistered. As one rheumatologist later noted, "It's an understatement to say that Dr. Polley was relieved when that morning was over!"[19] Years later, when the charade was revealed, Hench's apology for the deception was "accepted graciously" by the Danish visitor.

In fairness, Hench was not the only person agitating for secrecy in these trials. Merck & Company recognized the need to keep these studies confidential. An oft-repeated anecdote illustrates Merck's efforts in this regard. According to legend, at the same time the Mayo group got their original samples of compound E from Merck, small aliquots were also supposed to be sent by Merck to Dr. George Thorn in Boston and Dr. Robert Loeb at New York University. These investigators had requested samples to perform their own studies on the drug. Dr. Carlisle reportedly let slip to Dr. Polley that the shipments of compound E had either been "mysteriously" lost or had somehow been taken back from the investigators to whom they'd been sent. As a result, the Mayo group was the only entity that had access to existing supplies.

But despite the best attempts at secrecy, there will always be those who suffer from congenitally loose lips. They simply cannot keep a secret. Polley described one near disaster that nearly blew the cover off the covert project.

In February 1949, Dr. Hench and I were standing in the doorway [of Room 119, Saint Marys Hospital] one morning about 8:30 or 9:00 A.M. when Dr. Waltman Walters . . . [the same Dr. Walters who may have been the subject of Walter Alvarez's "nepotism" comments after his son, Luis, had complications following gall bladder surgery] came by with a retinue of sixteen men [I counted them]. As he passed by the doorway, Dr. Walters said rather loudly to Dr. Hench, "Good Morning. How are you coming with compound E?" We were standing on a terrazzo floor, but Dr. Hench nearly went through the floor anyway and responded meekly with "fine." Dr. Walters was a member of the Mayo Board of Governors at the time and we thought that he would have been sufficiently aware of the need not to say publicly what was going on, but apparently none of the accompanying men heard what the conversation was or understood it or cared about it as no sequelae were later encountered.[20]

The constant attention demanded by the compound E trials was a burden for everyone. It affected the rheumatologists' and chemists' workday, and perhaps more important, it also affected their home life. Even the investigators' wives felt the stress. Chuck Slocumb's wife, Mary, humorously brought attention to this fact in a most artistic manner.

Mary Slocumb was a talented artist, and on one of the numerous nights when her husband was detained late at the hospital, her loneliness got the best of her. Feeling neglected, Mary spent much of the evening painting a self-portrait, which she entitled "Me and E." The portrait showed her dressed in the convent clothing worn at that time by the Sisters of Saint Francis working at Saint Marys Hospital. Normally a conservative, proper woman, she later became "overly self-conscious" about the not-so-subtle sexual implications of her work—although it is clear that her friends thought the painting was socially acceptable. And hilarious.

CHAPTER 24

Convincing the Skeptics

There is nothing like firsthand evidence.

—SHERLOCK HOLMES, *A STUDY IN SCARLET*

As the trial progressed, more patients with rheumatoid arthritis in Rochester received treatment with compound E. And the results continued to be spectacular. Maybe too spectacular? As the number of patients successfully treated with compound E began to approach twenty, Merck & Company became nervous. As Dr. Polley noted: "In March, 1949, Merck & Company insisted on trials of compound E in patients in other parts of the country prior to any announcement of our preliminary results. Although it was never verbalized, as far as we knew, it was quite evident that there was concern about the 'Mayo mystique.' . . ." The company also wanted to be sure the world understood, ". . . the thoroughness with which Merck & Company conducted their role in the studies."[1]

The New Jersey pharmaceutical giant needed to be certain that the results reported by Hench and crew were reliable. Was compound E really the miracle drug it appeared to be? Or was there something magical in Rochester's water supply that temporarily cured these patients? Or something different about patients with rheumatoid arthritis in the upper Midwest? Or were the Mayo doctors somehow perpetrating a giant hoax? At this point, due diligence required the involvement of some non-Mayo participants. Dr. Hench was therefore asked by Merck to coordinate a multisite trial.

Hench arranged a meeting in New York City at the Waldorf Astoria Hotel. He contacted five of the country's most prominent clinicians and rheumatology experts. They included Drs. Edward Boland, a leading rheumatic disease investigator from Los Angeles; Paul Holbrook, a respected clinician from Tucson, Arizona; Edward Rosenberg, a clinician from Chicago and the man who had originally referred Mrs. G.'s roommate to Mayo; and Walter Bauer, a serious (to the point of being considered dour) leader in the field of rheumatic diseases from Harvard. A fifth physician, Dr. Richard Freyberg, director of the Division of Rheumatic Disease at the Hospital for Special Surgery and Chief of the Arthritis Clinic at New York Hospital, was, ironically, giving an out-of-town lecture when the assembly took place. He couldn't attend the meeting—even though it was taking place only a few blocks from his office. The invitees agreed in advance that they would participate in the study of a new antirheumatoid therapy, the identity of which wasn't going to be revealed until the New York meeting. It was no accident that the five clinicians chosen by Hench represented different geographical regions of the United States—after all, Merck wanted support for compound E to have a national base.

Freyberg, who was president of the American Rheumatism Association at that time, was crucial for the credibility of this group. Shortly before his death in 1999, he recalled with humor how aggressively Hench recruited him to join the others:

> Arranging the . . . meeting on short notice to bring together five busy academicians and clinicians was difficult. Fortunately Dr. Hench was a close friend to each. I recall that he phoned me at the 1948 meeting of the Canadian Medical and Chiurgical Society in Montreal and so sincerely requested that I come to the phone that the meeting host came to the podium, interrupted my presentation, and urged that I speak to Dr. Hench. I said, "But I'm in the middle of this presentation." He replied, "That's all right, I'll just tell them that you've been called away temporarily for an emergency call. I've got a few jokes that should hold them for a while." How could any of us argue with Dr. Hench when he was so persuasive?[2]

The New York meeting took place, but not without a few minor hitches. Dr. Rosenburg arrived from Chicago an hour late; he'd forgotten to reset his watch to eastern time. As already noted, Dr. Freyberg missed the meeting entirely because he was lecturing in Ontario that day. Hench eventually collected his four friends and sat them down in the hotel room. He described how his team at Mayo had been working on a highly potent antirheumatic therapy and that he believed they had made a major breakthrough. Before he revealed it to the group, he wondered if they would be willing to guess its identity. He pulled some hotel stationery out of the drawer and passed it around. "Go ahead, write down your guess," he challenged them. No answer won a cigar that day. The closest guess was "something to do with jaundice." The farthest, but certainly most honest, was a simple "I don't know." Hench broke into a grin as he read the answers. Then, in classic Hench style, he launched into a hyperkinetic description of compound E and the mind-boggling preliminary results the Mayo group had obtained. He took special pleasure in telling Dr. Rosenberg about Mrs. G., with whom the Chicago doctor's recently referred (and obviously close-mouthed) patient had been a roommate during the miraculous first use of the drug.

Hench carefully explained the situation with regard to Merck's concern. He offered a proposal, which was immediately accepted by all. The four doctors, along with the AWOL Dr. Freyberg, would come to Rochester and obtain "firsthand evidence" by observing the actions of compound E on patients with rheumatoid arthritis. After this, they'd be given samples of compound E to test in their own practices back home.

On the last Sunday in February 1949, the five master clinicians arrived from various parts of the country and converged on the Hench house. Hench had arranged a weeklong "experience like no other" in which the visiting doctors would participate in an intense compound E experience. Hench "turned his home into a clinic, complete with x-rays, movie pictures of earlier treatments, and posters of his studies." A microscope was set up on the dining room table; the doctors used it to review punch biopsies of the joints from patients treated with compound E. The Hench house was large and spacious, and his guests were encouraged to stay there with him during their visit to Rochester. There was room for everyone, and only a few of the visitors chose to sleep at the Kahler Hotel after spending their day at Hench's home.

The highlight of the week's activity was a meticulous study of two volunteers with severe arthritis. Starting Monday morning, the five doctors simultaneously examined these patients. Each doctor independently recorded his findings in a journal, which was updated daily. There was no sharing of thoughts or observations. After the initial Monday morning assessments, each patient received a huge intramuscular injection of compound E (roughly 300 milligrams). For the next five days the visiting doctors examined the subjects, "pounding on these patient's [*sic*] backs, prodding their joints, and testing their extremities at the same time. Despite this near assault, the patients cooperated beautifully."[3] What the doctors witnessed was amazing. "During the course of two days, we watched them miraculously improve."

The first day there was a generalized effect that caused one of the two test subjects to comment, "I've never felt so well during the time I've had arthritis as I do now." Over the next two days objective improvement in joint inflammation occurred. Prior to treatment, one patient had been unable to sit in a chair without collapsing into it—a move later called the "Hench flop" by diagnosticians.[4] By the second day he was rising and sitting in a chair unassisted, without resorting to the Hench flop. The other patient, a retired sea captain, had been unable to shave for years, so he'd grown a beard. On the second day after his injection, he shaved himself.

Unlike the other trials with compound E, this was literally a "one-shot" deal. The effects of the compound E bolus were expected to wear off quickly. And they did. By the fourth and fifth days, the dramatic improvements quickly disappeared, as did the patient's feeling of good health. A visiting doctor noted, "In one short week, we had seen the complete cycle of treatment response and realized that the positive impact of compound E would require prolonged treatment."[5]

When they weren't examining the patients, the swarm of doctors congregated at the Hench house and compared notes on various rheumatology topics. Hench treated the group like family, and they quickly became quite intimate with one another. Meals were prepared at the Kahler Hotel and delivered to the house, or Mary Hench would throw together sandwiches when between-meal snacks were needed. No one really took time to stop and eat. Years later one of the doctors who'd stayed with the Henches marveled at the integration achieved between the family and the physicians in such a short time. "To show how close we grew during that week, I can still remember watching Dr. Hench's son half-walking, half-crawling up the stairs to bed late one evening, his bare tail hanging out of his pajamas with his posterior flap down and his mother in hot pursuit."[6]

There were, of course, the occasional spats that occur between friends any time strong egos tackle big problems. Dr. Bauer from Harvard tended to be a little more belligerent than the others, and Hench wasn't one to back down easily—especially when compound E was involved. "We spent the evenings up at Phil's house with the microscope on the dining room table looking at biopsy specimens and Walter Bauer and Phil Hench were notorious for their heated arguments whenever they got together . . . friendly but volatile and understandable 'blue smoke' arguments for which Drs. Hench and Bauer were so well-known." Notwithstanding the occasional friendly conflict, "It was a week of miraculous discovery, exchange of information, and collegiality."[7]

At the completion of the Rochester trial, Merck & Company gave each of the five doctors enough compound E to treat two of their own patients back in their respective practices. It was also suggested that they test it on patients with forms of arthritis other than rheumatoid disease.[8]

The proof that compound E worked even when administered by non-Mayo doctors didn't come a moment too soon. Secrecy surrounding the compound E experiments had been maintained as long as possible; it would soon be necessary to announce or publish the results of this work:

> Merck . . . advised us at this time that it would not be possible to control the isolation of the studies for more than a six month period. . . . We were told that Merck had been cornering the world's supply of bile during this time and had a two or three year stockpile of bile, which of course was a cholesterol base that was used at that time for production of compound E. I (Howard Polley) used to wonder, perhaps inappropriately, still nonetheless, if the fact that they couldn't control the market anymore was really the reason we had to go to press.[9]

Dr. Kendall was out of town during most of the "clinical five" weeklong meeting in Rochester. He returned on Thursday and had an opportunity to meet the visiting doctors.

Kendall was described by one of them as "a delightful man, perhaps one of the most modest men I've ever met. Despite his many accomplishments, he took little credit for any of them. Mayo Clinic was so fortunate to have both Drs. Hench and Kendall. I don't know of any two men who worked as closely together as they did."[10]

This observation was on target—continued teamwork would obviously remain critical for the eventual success of the project. Unfortunately, as the study of compound E moved into the clinical arena, the chemist's role was beginning to diminish. It was crucial for team unity that Hench and his comrades keep Kendall and his achievements in the spotlight. There were still a few residual hard feelings at Mayo regarding Kendall's failure to have received "proper" credit for the isolation of thyroxin thirty years earlier. A few old-timers went so far as to insist belligerently that his achievement had been hijacked by others. To help ensure that any claim to this new treatment was secure, a series of dramatic before-and-after movies of patients receiving compound E treatment were shot. As Ollie Barnes, chairman of the Mayo Clinic board of governors at the time, pointed out, "They can't take this one away from us—we got it on film."[11]

CHAPTER 25

Announcement

There is so much red tape in these matters.

—SHERLOCK HOLMES,
THE ADVENTURE OF THE MISSING THREE-QUARTER

DESPITE THE BEST OF EFFORTS TO KEEP IT SECRET, RUMORS THAT the Mayo Clinic had made a breakthrough in the treatment of arthritis were beginning to circulate nationally. The preliminary compound E results had to be announced to the public soon or the story was certain to leak on its own. Hench arranged to make a presentation at the Association of American Physicians on May 3, 1949; this would serve to notify the scientific community of the results. But clinic protocol dictated that the Mayo staff should be "privately" notified prior to any public statement.[1] Early in April it was decided that the regularly scheduled staff meeting of April 20, 1949, would be the appropriate time and the place for this initial presentation. Although the topics for the weekly staff meetings were never announced in advance (because it was assumed that everyone would attend whether or not they knew the program in advance), word quickly spread that this meeting would be unlike any other in recent memory. Nervous energy continued to build during the weeks leading up to the highly anticipated event.

Dr. Clarence Kemper had just finished his Mayo Clinic Fellowship in rheumatology. As per Mayo custom at that time, he was making short visits to centers of rheumatology excellence around the country in order to ensure that his training hadn't omitted anything of importance. On April 19, 1949, Dr. Kemper was visiting a rehabilitation center in New York City.[2] The center was run by Dr. Howard Rusk, the world's preeminent physiatrist and a leading expert in arthritis. Kemper felt lucky to have his visit hosted by the famous physician. Imagine how the young doctor's various sphincters must have tightened when, early during afternoon hospital rounds, Dr. Rusk leaned over and asked him, "When is Mayo's new arthritis discovery going to be announced?"

In the words of Dr. Polley, "Our trainee was understandably surprised and nonplussed . . . he countered with, 'What do you know about it?'" Rusk admitted hearing only vague rumors; he wondered if the stories were real, and to what extent this "breakthrough" had been exaggerated. There was only one piece of information the young Mayo graduate could pass along to Dr. Rusk.[3] Kemper smiled back at the senior physician and said, "As a matter of fact it's going to be presented tomorrow night."

Kemper had no way of knowing, but his seemingly harmless confirmation of the pending announcement set a frantic cascade of journalistic events into motion.

Dr. Howard Rusk was not just a physiatrist; he was also a writer and an associate editor for the *New York Times*. As much as he would have liked to cover this breaking story himself, the timing of the meeting, just over twenty-four hours away, made it impossible. What else could he do? Rusk attempted to contact his boss at the *Times*, science editor William Laurence, by telephone; he was told that Laurence was attending a meeting in Detroit. With a sense of urgency he told the secretary, "Get in touch with him, tell him to take the next plane to Rochester, Minnesota, and call me when he gets there."

Twenty minutes later Laurence got the cryptic message from Rusk; a waiter hand-delivered it to the editor as he sat at a dinner table in a posh Detroit hotel. Laurence realized he had just enough time to get to the airport and catch the last plane to Rochester. To make his flight, he left directly from the restaurant without returning to his room. His luggage, along with all of his clothing, remained behind.

William Laurence's unexpected arrival in Rochester on the morning of April 20, 1949, created an immediate headache for the Mayo Clinic. The announcement regarding compound E wasn't scheduled until that evening, but Laurence, the consummate reporter, was already attempting to identify and interview the principals involved in the discovery. Worse, he was asking for an advance copy of the manuscript and/or any remarks to be made that evening. The Mayo Clinic administrative staff, who were justifiably puzzled as to how Laurence had found out about the meeting in the first place, considered their options. Advance copy? They told him "no."

Mayo, still maintaining strict secrecy, had already decided not to notify the local press about the announcement that evening. Saying "no" to William Laurence should have been easy. But William Laurence was not your average journalist. He was possibly the most famous science writer in the country at the time.[4] Mayo's decision to deny Laurence an advance copy of the announcement was not just tough; it was, in the words of some, "agony." But the denial was in keeping with policy, and if there is anything the Mayo Clinic does well, it's "adhere to policy." "No advance manuscripts" was the rule, and, as always, the rules would be followed to the letter.

Laurence considered this rebuff "an affront to his integrity." He also considered the inflexible Minnesotans who ran Mayo to be, at best, naive midwesterners—and, at worst, bona fide hicks. The clinic's leadership, perhaps recognizing exactly how bad Laurence could make them look nationally, offered a small olive branch. They would allow him to attend the staff meeting that evening and thus be the only person from the press to do so. It was a striking compromise by Mayo standards, although Laurence never realized or appreciated how the clinic's rigid tradition was bent to accommodate him. Indeed, according to Polley, he still felt Mayo had slighted him, and it lingered as an insult that "he never forgot . . . as long as he lived."

Kendall never forgot the events of that night, either. With William Laurence sitting in the front row scribbling notes furiously, a well-prepared Philip Hench walked to the front of the room, received an appropriate brief introduction, and began his presentation. Kendall later recalled:

> It was a memorable meeting. Every member of the staff of the Mayo Clinic and Mayo Foundation who could be there was present. Every seat in Plummer Hall was taken, and chairs were placed in the aisles, but many sat on the window sills and even around the platform. Others stood along the sides of the hall and even out to the elevators. Everyone was filled with anticipation and enthusiasm. Seldom has an advance in clinical medicine been made known under more auspicious circumstances.[5]

There was still a little room for levity. Compound E would be the first discussion of the evening; the final talk was scheduled to be a relatively low-key case report describing a rare complication of uterine fibroids. A senior pathologist acting as the master of ceremonies that night stood to introduce the speakers. He raised his arms to the physicians packing the room, then nodded politely to William Laurence. "I'm happy to see that so many of you are interested in uterine tumors," he said with his usual deadpan expression.

With these words and the polite laughter that followed them, Hench rose to the podium and began to describe the work his team had performed. According to Kendall, Hench "gave a brief review of the history of rheumatoid arthritis and of his own contributions in regard to the effects of jaundice, pregnancy, and the reversibility of the disease. He pointed out the possibility that reversal of symptoms might be brought about by 'substance X.' Was 'substance X' a hormone of the adrenal cortex?" he asked. Hench followed with an overview of Kendall's discovery of compound E, and described the work by Merck & Company to make it available for study. Near the end of his presentation, he explained the delicate logic behind the decision to administer compound E to patients with rheumatoid arthritis.

Then came the movies. A film record—obtained before and after compound E administration—had been made of all fourteen patients described that evening. The room was stunned. Kendall later recalled the impact of the moment: "After Dr. Hench presented his paper, the motion-picture film was shown. This illustrated the remarkable changes that had occurred in some of the patients. Only a phlegmatic person can watch that film without a lump in his throat or a mist over his eyes. For those who had known and worked with the patients it was a source of deep emotion."[6] One member of the audience, still flushed with emotion as he recalled the moment over 50 years later, noted that "It was like God had touched them."[7]

Hench's presentation was over in twenty minutes. Now came Kendall's turn to talk. "As I walked to the speaker's desk after these two superlative presentations I had a strong inclination to simply say that anything that I could add would be an anticlimax, and sit down."[8] But he didn't. Realizing that this might be his big payoff for nearly twenty years of research on the hormones of the adrenal cortex, he wanted to make sure the audience understood that "pure research in biochemistry sometimes can supply a vital link for use by the clinician." The chemist spoke eloquently about his work for fifteen minutes.

When he finished, Kendall returned to his seat. The first presentation on the effects of compound E in patients with rheumatoid arthritis was over. How had it been received? "The applause that broke loose immediately after the conclusion of the program has never been equalled at any other meeting that I have attended," recalled Kendall.

William Laurence of the *New York Times* wrote "a competent, accurate story" detailing the findings for the world; if he'd been permanently offended, he swallowed his anger for one night and helped celebrate a great moment in science. His unflinching journalistic integrity would not go unnoticed. Later that year Laurence won the prestigious Lasker Award for Medical Journalism as recognition for his superb coverage of this event.[9]

Nobody was surprised when Merck suddenly faced a huge demand for compound E. The new drug was an instant international sensation. Overnight, requests for the material came pouring in, and Merck found itself unable to appropriately assess the relative merits of each request. How should the company distribute its limited supply? Using a model that had been developed years earlier for the distribution of penicillin (when it was a rare, desperately sought new drug), Merck turned to the National Academy of Sciences to oversee compound E distribution until adequate supplies became available. The academy formed a committee, which subsequently evaluated all requests and disbursed the limited drug supply accordingly. This approach created a sense of fairness.

Now a bigger decision was looming for both Kendall and the Mayo Clinic. It would soon be possible to manufacture large quantities of compound E—providing that the various parties involved in its production could reach an agreement. To do this an armada of red tape had to be overcome. Five separate entities held patents related to the production of cortisone; these included Merck & Company, Ciba, Organon, Schering, and Kendall himself. Kendall and Merck & Company appeared to hold the key patents. Some of the older, more general patents, such as one that prohibited the use of any steroid that had a ketone group at carbon 11 and an atom of bromine at carbon 12, were insufficient to allow the patent holder to manufacture compound E, but could conceivably be used to block the manufacture and sale of this substance by others. As Kendall noted, "It was necessary to make some working arrangement, and this had to be done without delay."[10]

It should be clear by now that Edward "Nick" Kendall was a quirky individual with different facets of his persona that seemed, at times, to contradict one another. When it came to science, he typically overanalyzed things, usually settling on the hardest, most complicated, or least direct approach to a given problem. In short, his scientific thought process was both complex and thorough. On the other hand, he possessed a social and business naïveté that was almost embarrassing. Nowhere did this dichotomy become more obvious than in his thoughts regarding the patent conflict that was holding up the commercial production of compound E. Kendall's overly simplistic suggestion for resolution was to have every company with a vested interest assign its patent(s) to a third party.

What was needed was an independent party to bring the five (special interest groups) into agreement. This situation was one ready-made for the Research Corporation of New York. . . . The Research Corporation was given authority to grant licenses to each of the four manufacturing companies originally concerned and to any other company that complied with the terms of the agreement. Each of the contracting parties could use any of the patents held by the other members of the group. A small royalty was collected from each member that sold (compound E), and this was distributed among all members of the agreement.[11]

What would motivate a for-profit company to participate in an agreement of this sort? To anyone familiar with modern pharmaceutical manufacturers, it seems obvious that the various participants reluctantly agreed to cooperate in order to achieve their primary business objective—to make compound E. And, by doing so, make a lot of money. But according to Kendall, the driving force behind the agreement was something far more altruistic. The chemist sincerely believed that the various companies and patent holders were cooperating because it was the right thing to do. "One desire," he wrote, "was held in common by all the manufacturing companies: None of them wished to have a monopoly on the sale of (compound E)."[12]

Was Kendall really so naive as to believe that no drug company wanted exclusive rights to the most sought-after pharmaceutical product in history? That's a tough pill to swallow.[13] But Kendall's own motives were beyond dispute, because what he did next was extraordinary—and probably something Russell Marker would have never understood.

Patents are issued only to those who make the invention, but in many cases a patent is owned by the company or institution that supported the work and provided the necessary capital. Patents were granted to me and other members of the laboratory as each new step in the preparation of (compound E) was devised. The Mayo Clinic did not propose to exploit these patents for financial gain. Yet the Mayo Clinic did wish to be in a position to prevent any use of the scientific contributions made by members of its staff that would be contrary to the principles of the Mayo Clinic and of the medical profession. As soon as the agreement had been prepared, I assigned all my interest in the patents to the Mayo Clinic, which in turn gave them unconditionally to the Research Corporation.[14]

Was this pure unselfishness, or just more naïveté on Kendall's part? It was likely Kendall was simply doing what the clinic—and his own sense of propriety—demanded of him. His stated belief was that "no physician engaged in the practice of medicine should profit from the exploitation of any drug, vaccine, or appliance used in the practice of medicine. This is a time honored statement; it has been the policy of the Mayo Clinic from its beginning."[15] But it was not necessarily a belief that had always been intrinsic to Kendall; this was a principle he had learned, and he had his boss, Dr. William Mayo, to thank for instilling it in him. As he recalled from a similar situation many years earlier, when he isolated the thyroid hormone: "The method for isolation of thyroxin was patented, but when the patent was issued, Dr. W. J. Mayo offered it to the Board of the Trustees of the American Medical Association. The gift of the patent was made unconditionally." Kendall notes that "the incident made a strong impression on me. What was the best policy for the Mayo Clinic in 1920 would still be the best policy in 1950, in so far as it concerned financial return from patents in medical affairs." Kendall realized the potential for money to cause conflict with his partners; many had made significant contributions to the isolation and identification of compound E, but would be unlikely to benefit financially from the patents. "If I received financial reward, should I divide it among my colleagues? If I did not divide it, would this callous behavior produce bitterness and division?" Kendall realized that "in the long run teamwork is of more value than solo performances."[16]

It seems that Dr. Kendall was not totally clueless about real-world financial matters, but rather a man who believed deeply in the value of scientific charity. Nor was he so sophomoric as to think he would always hold this position. "Perhaps public opinion and medical practice will change in the matter of private gain from patented discoveries, and perhaps present customs will be revised. Until that time, I believe that all scientists in the medical institution should be willing to conform to established policy. I did not regret my choice of action."[17]

Dr. Kendall never seemed to regret any of his decisions or actions. And in this case, one suspects, neither did the drug companies that benefited by his noble stand—after all, they went on to make billions from his discovery.

CHAPTER 26

The Prize

Watson here will tell you that I never can resist a touch of the dramatic.

—SHERLOCK HOLMES, *THE NAVAL TREATY*

IN 2006 A HISTORICAL VIGNETTE ABOUT WALTER ALVAREZ WAS PUB-lished in the *Mayo Clinic Proceedings*; it mentions that John F. Kennedy spent one month at Mayo undergoing an evaluation for Addison's disease—and that he "had lunch" with Dr. Alvarez.[1] If so, Kennedy must have been among the last patients that Alvarez saw as a Mayo physician. The renaissance gastroenterologist retired from the clinic at about this time and moved to Chicago, where he signed on as a medical columnist for the Chicago Tribune Syndicate.

Soon known as "America's family doctor," Alvarez became famous; his newspaper column was read by millions. He also wrote numerous books and made regular appearances on television and radio. Alvarez's success as a public educator was a stunning career-capper for a man who had already excelled in both research and patient care. The Mayo Clinic logo consists of three interlocked shields, which are said to represent research, patient care, and education. True to form, Walter Alvarez was now the undisputed master of all three "shields"—he'd hit the academic trifecta.

Things were going even better for his son Luis. When the war ended, young doctor Alvarez returned to the University of California at Berkeley and was quickly promoted to full professor. The rapidly maturing scientist changed his focus of interest again, this time to the field of high-energy particle physics. He began designing and constructing particle accelerators that could propel electrons and protons at high velocity. By 1947 Luis had an operational forty-foot-long proton accelerator.[2] But an even more interesting physics device lay in his future.

It didn't take someone like Sherlock Holmes to deduce that the response to Mayo's announcement on compound E would be big. News of the breakthrough spread contagiously through the scientific community. Within days Kendall and Hench were invited to address the prestigious National Academy of Sciences. It was only the first of many major speaking invitations to come.

The two scientists decided to change the name of their discovery. Until now it had been referred to as "compound E" because it was the fifth substance Kendall had isolated from the adrenal cortex (the first four had been called A through D). But as the nondescriptive name hit the media, it immediately became confused with more familiar entities like vitamin E.[3] A better moniker was needed. In May 1949 Hench came to Kendall's laboratory and suggested that they come up with a "more distinctive name."

It was normally up to the pharmaceutical company to invent snappy, marketable names for new drugs, but in this case Hench and Kendall opted to do it themselves. Since they had been searching for "cortin," it made sense that the word they chose would conjure up the notion of cortin. Kendall suggested "corticosterone," and then removed the "ticoster" to leave the word "corsone." Hench balked at this; to a clinician, anything that began with the syllable "cor" implied something cardiac in nature. According to Kendall, Hench suggested adding the letters "ti" back into the middle of the word, creating "cortisone."[4] That account of the episode sits well with the recollections of Dr. Hench's junior colleague, Howard Polley. In his unpublished notes, Polley disputes a once-popular theory of how the word "cortisone" came into existence—the idea that it was a shortened version of the drug's chemical name. "The letters in the word cortisone are easily identified in the chemical name, 17-hydroxyl-11-dehydrocorticosterone," Polley notes, "but Dr. Hench had so much trouble with this chemical name that it's problematical that he used that chemical word to generate the term cortisone." In other words, Polley doubts that his boss simply shortened a word that, given his speech impediment, he would have found unpronounceable in the first place.

Percy Julian was still working at the Glidden Company when news about cortisone grabbed the nation's attention. As the price of cortisone quickly rose to over $4,000 an ounce, Julian realized that a more economical way of manufacturing the drug was necessary. Sarett and his colleagues at Merck had made the commercial production of cortisone possible. Now someone needed to make it practical.

Julian had a sudden flash of chemical inspiration. Rather than trying to synthesize cortisone (Kendall's compound E), the chemist realized that he could create similar substances from soybean derivatives. It took him only a few months to synthesize Reichstein's substance S, also known as cortexolone.[5] By adding an oxygen molecule to this compound he converted it to cortisol (or hydrocortisone). Within the next few years biochemists and microbiologists at the Upjohn Company in Michigan discovered a mold that could oxygenate compound S and transform it into a substance that was easily converted to cortisone. This was a huge breakthrough. Not only did Julian's various new synthetic pathways offer fresh ways to make cortisone at a lower price, but they also made a variety of "cortisone-like" compounds available (such as cortisol, hydrocortisone, and so forth), each with slightly different actions and side effects. Other companies could now competitively develop and market their own unique, patented drugs. The stage was set for the cost of cortisone-like substances to drop dramatically.[6]

The discovery at Upjohn of a microbiological process for converting compound S into other steroids was a major achievement; it meant Julian could transform soybean products into all sorts of new wonder drugs, including cortisone. But as good as this breakthrough seemed, it got even better when it was realized that the same microbial process could also metabolize progesterone into cortisone. Substance S was a good synthetic starting material from which to make cortisone, but progesterone was even better. And thanks to Russell Marker, Mexican yam-derived progesterone was now being produced in massive quantities by Syntex and other companies.[7] The female hormone made a cheap, efficient substrate upon which to base a cortisone-manufacturing industry.

Julian brought his observations to the leadership at Glidden; he proposed that the company quit manufacturing cortisone from soybeans and instead switch to making it (and similar compounds) from Mexican yams.[8] The reduced cost and improved efficiency would likely make Glidden the leader in the field of cortisone production.

Glidden had other ideas. It was a paint company—always had been, probably always would be. Julian's steroid research, while exciting, was distinctly outside its corporate comfort zone.[9] Unknown to Julian, the corporation had decided to divest itself of steroids entirely. It had already sold off the steroid-manufacturing operation to Pfizer, and Julian's grand plans were met with orders to teach his various processes to the chemists at Pfizer. Help the opposition? No way. Russell Marker had merely been rejected by Parke-Davis and other companies when he proposed the creation of a steroid-manufacturing empire. Percy Julian, in contrast, was not just facing rejection—he was being asked to cooperate with the dismantling of his dreams. For Julian, this was too much to endure. Like his rival, Russell Marker, if he wanted to make steroids, he'd need to do it on his own.

Percy Julian left Glidden immediately and set up Julian Laboratories. His first facility, located in Chicago, was a quickly converted, rat-infested, dilapidated warehouse. Despite the humble surroundings, Julian successfully manufactured a variety of corticosteroids from soybeans, using processes with which he was very familiar. Companies like Upjohn, Ciba, Pfizer, and even Merck bought his products, especially progesterone.

But Julian could see the writing on the wall. Steroids made from soybeans were good, but those produced by Syntex in Mexico, using the Mexican yams, were just as good as—and far cheaper than—his own. They would, as far as the chemist could see, always be cheaper. If he wanted to continue competing in this market, he needed to switch his starting material to yams. That turned out to be an easy decision to make, but a tough one to implement. Julian was short on capital, and few banks would make loans to blacks at that time. By cobbling together a consortium of private investors and friends—and by using his own ability to borrow money—he was eventually able to open Laboratorios de Julian de Mexico near Mexico City.[10] With his new facility, Julian was ready to reenter the competitive field of steroid production.

Or he would have been. But there was a *problema politica*. For reasons that later became obvious, the Mexican government refused to allow Julian to harvest the yams he needed. Without these yams, Syntex and the other Mexican proprietors enjoyed a monopoly on the production process.

Was Julian defeated? It seemed there was no option except to bail out of Mexico and return to the United States. However, at this point a man named Abraham Zlotnik made Julian's reacquaintance; years earlier Julian had cashed in many of the connections and favors he'd acquired while studying in Austria to help Zlotnik escape from Nazi Germany. Now the grateful man wanted to return the favor. After learning of Julian's dilemma, Zlotnik offered a proposal. He was certain that the yams Julian needed grew in Guatemala as well as Mexico. He'd be delighted to obtain them for his former benefactor.[11]

Julian accepted the offer. Zlotnik, using his knowledge of Central American geography, quickly found a steady supply of the tubers in Guatemala and arranged to ship them to Mexico. With Zlotnik's help, Julian was soon back in business, producing progesterone and other steroid derivatives from his Guatemalan connection.

Julian never stopped improving his production methods. Eventually another chemistry breakthrough was made, and it became possible to quadruple the production of steroids from yams. Most companies would have used this boost to increase their profit margin. Not Julian. He insisted that the company reduce the price of progesterone, which was then $4,000 a kilo, to $400 per kilo.[12] The cost of other synthetic steroids fell accordingly. Percy Julian, like Russell Marker, wanted to make money. But he also wanted to help people. He wasn't going to be responsible for limiting the miracle of cortisone, progesterone, and other steroids to the very wealthy.

The Mexican situation did not go unnoticed by the U.S. government. In the mid-1950s the U.S. Senate held public hearings regarding Syntex and the allegations that it had influenced the Mexican government to maintain a monopoly on the coveted yams. Julian's company was one of several that claimed damages, and his testimony made him the star witness at the Senate hearings.[13] As a result of these actions, pressure was placed on the Mexican government to make access to yams readily available to anyone who wanted to buy them. Julian was able to secure Mexican yams for his own production, and the company flourished.

In 1961 Julian sold his laboratory to Smith, Klein, and French for just under $2.5 million. Coupled with the money he had been earning as its president, Percy Julian was suddenly one of the richest black men in America. He retired from business and spent much of his remaining days as a public speaker. The National Academy of Sciences, not an organization to readily accept black members, offered him admittance in 1973. It was another milestone in Julian's career. Two years later Percy Julian was dead from the bane of all organic chemists—liver cancer,[14] likely caused by the oceans of dangerous solvents and chemicals to which he'd been exposed. Julian was still talking about chemistry, seemingly oblivious to the fact that he had a fatal condition, right up to the moment of his death.

On the evening of October 25, 1950, Nick Kendall's married daughter, who lived in New York State, received a telephone call from a newspaper correspondent.[15] He wanted to know about her father. What were his likes and dislikes? Did he have a favorite sport? What churches or clubs did he belong to? When she asked why the reporter would want to know these things, he swore her to secrecy. Apparently no more able to "resist a touch of the dramatic" than Sherlock Holmes, the anonymous voice informed her that "he had no doubt" the Nobel Prize in Physiology or Medicine was going to be awarded the next day to her father.

Stunned, she called her brother in Rochester, New York, unapologetically breaking her promise not to tell anyone about the phone call and asking her sibling for advice. "What shall I do?" she asked. "For God's sake hurry up and telephone your father," was her brother's reply.

It was from her subsequent call that Edward Kendall got his first inkling of the impending Nobel Prize. Keeping his hopes in check, he went through his normal activities on the following day—although he uncharacteristically left word with his secretary as to where he could be reached when he went to the dentist. Throughout the day other correspondents phoned him, but all carried the same message—no official vote had been taken yet. Kendall refused to talk with them.

At 1:30 P.M. on October 26, 1950, the official announcement was made. Edward C. Kendall, Philip S. Hench, and Tadeus Reichstein were the recipients of the Nobel Prize in Physiology or Medicine for their "investigations of the hormones of the adrenal cortex." It was official. It was public. And all hell was breaking loose.

Hench, the perpetual world traveler, was in Ireland when he received the news that he had won. He immediately cabled Drs. Polley and Slocumb and informed them that they would share in his prize money.[16]

CHAPTER 27

Stockholm

Life is infinitely stranger than anything
which the mind of man could invent.

—SHERLOCK HOLMES, *A CASE OF IDENTITY*

On October 26, 1950, it was announced that Edward Kendall and Philip Hench, along with Tadeus Reichstein, had won the Nobel Prize. Three days later, on October 29, King Gustav V—who would have normally bestowed the award on the new recipients—died. At ninety-two years of age, his death was not unexpected. He was immediately succeeded by his son, fifty-eight-year-old Gustav (VI) Adolf. The old king's death was an unfortunate beginning to the biggest event of which Kendall or Hench would ever be a part.

The Kendalls' trek to Sweden for the December awards ceremony began well. The chemist and his wife were taken from Rochester to Chicago aboard the private car of the vice president of the Chicago-Northwestern Railroad Company.[1] The traveling palace, complete with big beds, maid service, and freshly squeezed orange juice, made the initial 500 miles of the trip a pleasant adventure. After reaching the East Coast, they "crossed a well-behaved Atlantic Ocean on the steam ship *Oslofjord* to Copenhagen and Oslo." From there they took a train to Stockholm. Nick Kendall and his bride of thirty-five years arrived on December 8 and checked into the Grand Hotel.

Philip and Mary Hench also arrived in Stockholm by train that day. They weren't alone. In tow were their four children—Kahler, Mary, Susan, and little John—along with Philip Hench's mother-in-law. Howard Polley points out that "Phil took his family to Stockholm and got a lot of attention at the time of the Nobel Prize, because I guess it's about the first time anybody had ever done that." Hench's own recollections confirm the unusualness of a family entourage at this event.

> To our rooms came each day a small army of friendly newspaper folk who seemed to be fascinated by the fact that such a large family had journeyed so far to the Nobel festival. Two reporters were especially impressed by the inclusion of my wife's mother, and one newspaper headline read "Mother-in-Law attends Nobel Festival."[2]

Hench had harbored serious misgivings about the appropriateness of bringing his entire family. It turned out to be unfounded.

> The news folk and the various Swedish friends we met felt that we had paid them a real compliment in bringing the whole family. Soon our four children, especially 7-year-old John, and the charming, 17-year-old daughters of Dr. Reichstein and the American novelist, Mr. William Faulkner, became the delight and daily target of the reporters and news photographers, who understandingly became much more interested in these young people than in the older guests.[3]

The 1950 Nobel Prize Festival was an extra-special event; it was the fiftieth anniversary of the Nobel Prize. All 100 living Nobel laureates from the past had been invited, and about 25 of them were able to attend. As the Kendalls and the Henches awoke in the Grand Hotel the next morning, they discovered that the breakfast area was packed with former prizewinners—including Scotsman Alexander Fleming of penicillin fame, and Gerhard Domagk, the German discoverer of the first sulfa antibiotic. Mingled with them were ambassadors and diplomats from all the countries that had sent a past laureate. The American ambassador, William Butterworth, was present and accompanied by his wife. The entourage from Rochester ate their Swedish pancakes that morning with Mrs. Butterworth.

In the afternoon a reception for the contingency was held at the Nobel Foundation House. Tadeus Reichstein, the third member of the trio, was not there; a dense, impenetrable fog worthy of Stephen King surrounded Stockholm and would delay his arrival until just before the main ceremony on Sunday. In his absence Drs. Hench and Kendall were toasted by their hosts for their accomplishments. True to character, Hench rambled on in his response. In contrast, Kendall—according to the wife of the British ambassador—gave "the shortest speech" anyone associated with the ceremonies had ever heard.

The following day, Sunday, December 10, turned out to be, in Kendall's words, "long and arduous." This was the anniversary of Alfred Nobel's death in 1896, the day designated for the "solemn festival of the Nobel Foundation" that would take place in the city's Concert Hall between 4:00 and 6:30 P.M., and then move to the city hall from 7:00 P.M. until 2:00 A.M. The day's activities began mid-morning with the laying of wreaths on the grave of Alfred Nobel in Stockholm's North Cemetery. Despite the distractions of fog, melting snow, and soggy grounds, the somber memorial service proved to be a deeply emotional experience for the Americans.

Returning to town later in the morning, the participants began rehearsing for the main event that afternoon—the awarding of the Nobel prizes. Televisions were placed outside Concert Hall so that those unable to obtain seats could nonetheless watch the ceremony. While workmen moved the sets into position and began testing them, Hench noticed some of the former prizewinners stopping to marvel at the strange gray boxes. Television was still a novelty in 1950, and many Europeans had never seen one.

> As the laureates and their escorts entered, one of the television staff turned his camera on them. The distinguished laureates, like a group of American youngsters, became fascinated as they saw themselves and their companions on the screen. Most of the European laureates probably were seeing television for the first time. The laureates in physics, some of whom probably discovered some of the fundamental principles used in radio and television, appeared to be as fascinated as the rest.[4]

Hench admittedly floundered his way through the rehearsal.

> Apparently, some of the men were out of practice in bowing to royalty and it was amusing to see dignified men practicing nodding solemnly to an empty chair, often in too stilted or too vigorous a fashion at first, then in a more subtle, polished manner. . . . Sir Henry Dale was not satisfied with the short, unsophisticated nods which Dr. Kendall and I were making, for he came over to us, chuckled and said "For this one afternoon you two fellows will just have to lay aside your democratic principles, and really throw yourselves into this thing."[5]

Kendall's recollection of the preceremony preparations was, as usual, somewhat different from that of his much more flamboyant colleague.

After participating in the same rehearsal as Hench, the notorious minimizer merely noted that "our part was short and not complicated."[6]

A quick lunch followed, and everyone—including Tadeus Reichstein, who had finally beaten the fog and made it into Stockholm—began changing into formal attire. At 3:15 P.M. attachés from the Swedish Foreign Office arrived at the Grand Hotel to escort the honorees through the crowded streets of Stockholm to Concert Hall. It looked like a migration of blonde, blue-eyed Scandinavian penguins. Swarming around the slow-moving procession were the roughly 2,000 attendees who would soon fill Concert Hall, and each of the men—including the paparazzi—were decked out in traditional black-and-white formal attire. Another 3,500 people, many equally well dressed, moved into position to watch the ceremony on the various television screens in two nearby auditoriums. Concert Hall itself was decorated like the Garden of Eden with various plants and flowers that had been flown in from Holland.

At precisely 4 o'clock, with the auditorium packed and the various participants—minus the award recipients—arranged on the stage, a trumpet fanfare announced the entry of the Royal Family. The orchestra played "The King's Anthem" as the royal entourage took their seats. Because of the recent death of Gustav V, the women of the Royal Family wore black gowns with prominent white collars. Moments later a second fanfare summoned the new laureates into the hall. As Hench described it, "Two tall doors at the rear center of the platform were opened with ceremonial slowness by uniformed ushers. Then, as the orchestra summoned the king and the assemblage to their feet, the laureates marched onto the stage, stood for a moment in front of their seats, made their first reverence to the king, and sat down."[7]

Dr. Lars Birger Ekeberg, the Lord High Steward of the Nobel Foundation, opened the ceremony by welcoming the Royal Family and laureates, followed by a review of the life of Alfred Nobel and a tribute to the late king. In deference to the large local radio and television audience, Ekeberg spoke in Swedish, but printed programs containing his remarks were distributed in English, French, and German.

With the conclusion of Ekeberg's introduction, the conferring of prizes began. This occurred, by tradition, in the order in which they had been mentioned in Alfred Nobel's will: Physics first, next Chemistry, then the prize for Physiology or Medicine, followed lastly by the prize for Literature. With each presentation the king arose from his seat in the front row of the audience, stepped forward, and handed the prize to each recipient as their sponsor announced, "I now have the honor of asking you to accept the Nobel Prize for 1950 from the hand of his gracious majesty, the king."[8]

Hench later recalled his amazement when the king sat with the audience and not on the stage with the laureates and their "sponsors." He later learned that "it is traditional at Nobel festivals for the king and Royal Family to honor the laureates by leaving the stage to them. Rising from his seat placed just in front of the audience, the king welcomes each recipient and is 'thus the first of the people' to honor each prize winner."

The orchestra played in celebration each time a laureate received his diploma and gold medal from the hand of the king. A short, private conversation between Gustav Adolf and the laureate transpired during each prize transaction. The specific topics discussed during these brief conversations have never been ascertained; Hench once asked Kendall what the chemist and king had discussed. "With a twinkle in his eye, Dr. Kendall told me: 'That's a secret.' But I suspect that King Gustav Adolf asked Dr. Kendall: 'How do you pronounce 17-hydroxy-11-dehydrocorticosterone?'" And what of Hench's private conversation with the monarch? Perhaps the words exchanged between the king and Dr. Hench touched on the moment of panic occurring seconds earlier as the extremely large Rochester rheumatologist tripped on the edge of a carpet as he approached to receive his award. Hench would have looked less than elegant sprawled on the stage floor in front of His Majesty.

He had avoided embarrassment once, but Hench and the others still needed to leap one more hurdle without a faux pas. "After the awards were presented . . . one was supposed to walk backwards about 15 to 20 feet, keeping one's face to the Royal Family, until the steps leading back up to the platform were reached. The motion pictures later revealed clearly how successful or otherwise each recipient was in accomplishing this feat." Hench noted that, "Having seen the motion pictures, I must conclude that walking backward gracefully is not one of the things well taught to members of the staff of the Mayo Clinic. But from what I heard of the good humor and democratic instincts of King Gustav Adolf, I feel certain that he gave everyone an 'A' for effort."[9]

The Nobel Prize itself is a thing of physical beauty. It comes in two parts. The first is a diploma encased in an embossed, heavy, blue leather folder. The oversize sheepskin is illustrated with hand-colored drawings of various points of interest around Stockholm. The second part—the medal—is almost pure gold, and it's bestowed along with a blue hand-tooled leather case. On one side of the medal is a profile of Alfred Nobel, and on the reverse is an engraving of the "Spirit of Medicine," depicted as a figure holding an open book upon her knees while collecting spring water with which to "quench the thirst of a sick young maiden." The recipient's name is engraved onto the medal, along with a passage from Virgil's *Aeneid*; an approximate translation into English might be: "How pleasant it is to see human life enriched by the inventiveness of the arts."

After awarding the 1949 Nobel Prize in Literature (given belatedly) to William Faulkner, and the current 1950 Nobel Prize in Literature to Bertrand Russell, the ceremony closed with the entire audience singing the Swedish national anthem.

Following the ceremony at Concert Hall, approximately 1,000 of the luckiest attendees moved on to the banquet at Stockholm city hall. Held in the great Blue Hall with its high ceilings, cavernous hallways, and massive stairways, the crowd was well accommodated. The banquet that evening required seventy kitchen staff to prepare and 132 waiters and waitresses to serve. "A triumph in culinary logistics!" Hench proclaimed. The high point of the feast involved the presentation of dessert, which consisted of sculptured ice cream served on a tray with an accompanying statue made of clear ice. Each carving was internally lit by tiny, colored electric lights powered by small batteries, both of which had been frozen inside the ice. As the lights were dimmed and service began, the effect was one of a sparkling glacier sweeping majestically down the massive marble stairs from the kitchen to the banquet hall.

The banquet itself took four and a half hours and included numerous toasts. There were salutes to the new king, the old king, Alfred Nobel, the award recipients, and countless others. Interspersed between the toasts were short banquet speeches made by the new laureates. Hench, Kendall, and the still unsettled Reichstein were introduced by Robin Fahraeus of the Royal Academy of Sciences. In slow, simple, carefully practiced English, Fahraeus explained for the audience's benefit the significance of the awardees' work: "Together your researches have contributed to the enlightenment of the extremely complicated physiological chemistry of the suprarenal glands which since their discovery for a long time have been assumed to play no other part than to fill up the vacuum between the kidneys and the diaphragm."[10]

This colorful introduction was followed by equally colorful remarks from the new laureates. Hench, the most eloquent speaker despite his congenitally raspy, nasal intonation, thanked the many colleagues who had supported his work; the first name he mentioned was that of his close friend

Walter Alvarez. He admitted that the generosity of the Nobel Foundation and the Karolinska Institute "elicits an emotional response which perhaps I am physician enough to understand but which I am not artist enough to describe."[11] Hench went on to point out how the field of medicine needed to thank chemists like Kendall and Reichstein for their contributions: "Medicine . . . has been receiving its finest weapons from the hands of the chemist, and the chemist finds his richest reward as the fruits of his labor rescue countless thousands from the long shadows of the sick room."[12] It was powerful after-dinner rhetoric.

At 11:30 the thoroughly sated Swedes and their guests moved from the Blue Room to the even more magnificently decorated Gold Room of city hall. Dancing began at midnight and ran until 2:00 A.M., at which time all the guests, both honored and otherwise, were exhausted. Content at this point to understate the obvious, Kendall and Hench both commented that it had been a "memorable day."

In the days that followed there were more banquets, receptions, and speeches. Kendall and Hench felt and acted like royalty; abandoning the last vestiges of their plebeian midwestern attire, they packed their Homburgs and both rented high silk hats, which remained comically atop their heads for most of the next two days. Hench had several additional private conversations with King Gustav Adolf and was delighted to learn that the king had received an honorary degree from his alma mater, Lafayette College, during a visit to the United States in 1938.

Wednesday, December 13, was a special day in Sweden: the "crowning" of St. Lucia, the "Queen of Light" of Stockholm. The money raised by the festival that year was designated for treating children with rheumatic disease, and so it followed that the great American rheumatologist Philip Hench was asked to crown Lucia at the evening ceremony held, once again, at city hall. At the event Hench found himself face-to-face with a stunning Scandinavian blonde beauty, a woman he described as having "poise, charm, personality, and perfect features." She wore a crown of lighted candles and was escorted by a man holding a wet towel—which could be tossed over her head in the (not uncommon) event that her hair caught fire from the burning wax sticks. Hench had intended to conclude the crowning ceremony with a short speech, but he had developed "a very hoarse sore throat." With the approval of his host, Hench's eldest son, Kahler, gave his final speech that evening and "did very well indeed." As young Kahler completed his talk, Philip Hench hung a magnificent jewel pendant around Lucia's neck.

A young boy soprano began to sing as the crowd cheered. It was the last official duty any of the Americans would perform in Stockholm.

Soon the Kendalls and Henches, now mere tourists, returned to the United States aboard the ocean liner RMS *Queen Elizabeth*. Kendall, afraid of losing his Nobel medal, refused to pack it with his other belongings. He wore or carried it discreetly for the entire trip. The tired travelers arrived back in Rochester in early January 1951.

On the evening of January 12, the Mayo Clinic threw one final Nobel Prize–related reception at the Mayo Foundation House. It was another gala affair, attended by approximately 400 guests. Once again, Philip Hench brought most of his family, but this time he also generously invited the dozens of students, residents, fellows, technologists, and other junior members of his "team" who had made his Nobel experience possible.[13] It was his way of publicly saying "thank you" to all those who had helped him achieve this spotlight.

Nick Kendall also attended the reception that evening. Alone.

CHAPTER 28

Aftermath

*That hurts my pride, Watson. It is a petty
feeling, no doubt, but it hurts my pride.*

—SHERLOCK HOLMES, *THE FIVE ORANGE PIPS*

THE AFTERGLOW OF THE NOBEL PRIZE DOES NOT LAST FOREVER. THE
Mayo Clinic, and especially Drs. Hench and Kendall, had been basking in
its glory for nearly a year, but by the start of 1952 things were beginning to
return to normal in Rochester. The Mayo Clinic's leadership was in a period
of transition, and the new captains running the world-famous medical facility were steering it on a conservative downwind course. Perhaps the nautical
analogy is not such a bad one: the Mayo Foundation board now included
the chairman of Northwestern Mutual Life Insurance Company as one of its
new trustees.[1] Although his name meant little to the average Minnesota citi-
zen in 1952, Edmund Fitzgerald would eventually become famous because
of the legendary Great Lakes iron freighter named for him.[2] Another future
celebrity, Harry Blackmun, had left his Minneapolis law firm in 1950 and
joined the clinic as legal counsel. He was now serving on the Mayo Foundation board.[3] But not for long. On November 4, 1959, Blackmun would
be appointed to the U.S. Court of Appeals, and on June 9, 1970, he would
join the U.S. Supreme Court. Individuals like Fitzgerald, Blackmun, Kendall, and Hench speak to the depth of talent and leadership that was now
assembled in this small town in rural Minnesota, and their presence in early
1952 put the Mayo Clinic at the zenith of its fame.

Hench's life was returning to some semblance of post-Nobel normal-
ity. January 1952 found him reengaged in his "hobby" involving Walter
Reed and the story of yellow fever. More precisely, Hench was in Havana to

receive the prestigious Order of Finlay award from the Cuban government honoring his efforts to identify and restore Walter Reed's research headquarters.[4] Named for Carlos Finlay, the Havana doctor believed by many Cubans to be the real hero of the yellow fever story, the award was one of the most prestigious official recognitions the Cuban government offered—a Latin American version of the Nobel Prize in Physiology or Medicine. Hench's dedication to saving Camp Lazear—and, more important, the perception he'd fostered among Cubans that his work on the history of yellow fever was "fair and balanced"—had overcome the Cuban-American impasse that usually set Walter Reed's legacy against that of Dr. Finlay's.

The Havana media were enchanted by this charming, respectful, world-famous American with such an obvious affection for Cuba, its history, and the importance of Dr. Finlay's contributions.[5] Hench spent hours answering reporters' questions, most of which were directed at his plans to restore Camp Lazear through the combined efforts of the Cuban government and private donors. Sadly, Hench's consortium would eventually acquire ownership of Camp Lazear as planned, only to lose it forever following the Cuban Revolution.

Inevitably, reporters would ask Hench questions about cortisone. When one newspaper reporter queried him as to how cortisone could cure certain diseases, Hench explained:

> The new hormones cortisone (and ACTH) do not actually "cure" rheumatic fever or the disabling rheumatoid type of arthritis. . . . Instead they suppress markedly the inflammatory reactions in these and many other conditions, thus controlling symptoms.
>
> . . . These hormones are the most powerful ones ever discovered. Their effects are so striking and widespread that investigators have only begun to learn what we want . . . to know about them. So much remains to be learned, but much important information is being accumulated rapidly.
>
> Although these hormones still belong as much, if not more, to the physiologists and clinical investigators, than they do to the general practitioners and rheumatism specialists they are already most useful, indeed irreplaceable, in the management of a number of illnesses.[6]

Remarks like these foreshadow Hench's growing bias toward his discovery. Cortisone had only been in clinical use for a few years, but Hench was already enamored of its pharmacological potential. Some would say he was turning into a fanatical advocate for the drug. Yes, Hench recognized that cortisone had undesirable side effects: it increased blood sugar in an "anti-insulin" way, it depleted the body of certain minerals such as potassium, and with prolonged use it produced fatigue, muscle weakness, and numerous other problems. Hench periodically sought to "temper the enthusiasm of those who were likely to use cortisone as a harmless boon." His folksy aphorisms regarding cortisone were well known: "Cortisone is the fireman who puts out the fire, it is not the carpenter who rebuilds the damaged house."[7] Or how cortisone and similar drugs "were like an asbestos suit which enabled the patient to pass through the fire—the critical stages of the disease."[8] But there was no mistaking Hench's deeper sentiments. He believed that cortisone was a wonder drug poised to revolutionize the field of rheumatology. Many of his public statements regarding the safety and utility of steroid drugs may have seemed overtly conservative—"As a result of three years experience my colleagues and I at the Mayo Clinic and certain other American rheumatologists believe that cortisone and ACTH can be used safely and with great satisfaction in many cases of chronic rheumatoid arthritis"—but his private feelings about these drugs were becoming more widely known.[9] Hench thought they were miraculous therapeutic agents, and he was not about to have *his* drugs criticized. *By anyone.*[10]

His intolerance of any cortisone-related criticism struck those around Hench as something new and distinctly odd. Philip Hench had always been the personification of humor, the gentle voice of persuasion, the affable colleague, the comically kinetic cartoon caricature of a scientist. But Hench was changing, and the unsettling change seemed to be fueled by the widening controversy over cortisone.

John Hench was only seven years old when his father won the Nobel Prize in 1950, but he recognized his father's subtle behavioral transition. "I would say that my father did undergo something of a personality change and that it could have been a form of depression, though I doubt it was diagnosed at that time. Depression is not unknown in my family, nor in many other families."[11] Looking back, the prize may have accelerated his change, but perhaps did not precipitate it. As his youngest son also noted, "I am not able to pinpoint when his personality changed. I wouldn't say that it necessarily began right after he won the Prize, however."[12]

Hench's son watched the changes in his father's personality slowly—almost imperceptibly—evolve. "Considering how famous and busy he was and despite the fact that he was 47 years older than me, he was a very attentive and fun father before I went away to school. . . . I don't recall many real conflicts with him then." But the harmony at home began to change ever so slightly after his father became famous.

There were some unpleasant [conflicts] though, when I was older. Most of them had to do with his expectations for me. He had very clear ideas about what I should do with my life. I would go to Mercersburg Academy, where he and Mom had sent (my brother) Kahler. I did. I would go to Lafayette, where most male Henches went. I did, but he was not at all pleased when I seriously considered other alternatives. I would pledge Sigma Alpha Epsilon, his fraternity and Kahler's. I pledged Phi Gamma Delta. I always thought this was the most irrational of these insistences, since obviously any given house at Lafayette was vastly different 50 years later. I would become a physician. I did begin as a pre-med, but struggled with calculus and biochemistry. My biology professor diagnosed this as "pre-med syndrome"—deliberately or unconsciously doing poorly in order not to have to follow in your father's footsteps.[13]

What was driving this gradual change in Philip Hench's demeanor? His young son John had an inkling.

I was not in a position to figure it all out at the time, [but] I would guess that his personality change began with the mounting criticism of cortisone. Dad was himself very—perhaps in some cases blindly—loyal to friends, family and colleagues, and I'm sure that criticism coming from people he had counted as friends and colleagues, even if it may have been well-intentioned and appropriate, must have been hurtful. Along with other people . . . who don't draw boundaries between their work and the rest of their lives, dad found it very difficult to take criticism of his work and accomplishments as anything but disloyalty.[14]

If he were asked about the perceived criticism, Philip Hench—like his idol Sherlock Holmes—would have very likely admitted that his reaction to it was "a petty feeling, no doubt, but it hurts my pride."

Edward "Nick" Kendall was finished at Mayo. His career wasn't over because he wanted it to be over. It was over because the Mayo Clinic said it *had* to be over. In 1951 he hit the magic age of sixty-five. It didn't matter if you were a worker bee, CEO of the organization, or Nobel Prize winner—Mayo had a mandatory retirement age of sixty-five, and it wasn't breaking the rule for anyone. Not even Kendall. He received the proverbial gold watch, hearty handshake, and a quick escort to the front door. His laboratory was turned over to his subordinates, and with little fanfare the internationally famous Nobel Prize–winning chemist was suddenly unemployed.

Fortunately for Kendall, Princeton University had a different philosophy. A well-funded program had been established there to attract displaced or unhappy former Nobel Prize winners like Kendall. He was offered an office and some limited laboratory space. It was less than he was used to, but far better than nothing. Kendall accepted the offer and moved to the beautiful garden region of New Jersey. Considering the number of trips he'd made over the years to visit various pharmaceutical companies headquartered in and around Princeton—and to drop in on his chess-playing friend, Lew Sarrett—it may have been a less traumatic relocation than it seemed.

Within weeks of his move to Princeton Kendall was once again studying the adrenal cortex. It was widely recognized that cortisone and the other adrenal steroid compounds that had been identified so far all came with significant physiological limitations or side effects. Perhaps "better" adrenal hormones were still hidden within the complex cortex? Kendall had postulated as far back as 1945 that the adrenal gland ought to contain other substances.[15] Specifically, he suspected that one of the adrenal steroids— possibly a substance not yet discovered—might be chemically linked to vitamin C, the substance that Szent-Györgyi had initially discovered in the adrenal gland. If such a compound existed—and Kendall was sure it did— his gut instinct told him it would be a substance of major physiological importance. For the next twenty years, Kendall—the man who had isolated thyroxin and a host of adrenal corticosteroids—focused on isolating this new, mysterious compound.

But it didn't exist.

With each passing year there appears a new crop of Nobel laureates. The 1954 Nobel Prize in Literature went to Ernest Hemingway for his Cuba-set novel *The Old Man and the Sea*. With this award he joined the same elite club as Kendall and Hench, and doing so provided him with another ever-so-subtle new tie to the Mayo Clinic. But the Nobel Prize wouldn't be the connection to Mayo for which Hemingway would be remembered.

Hemingway didn't have the fairy tale Nobel experience shared by Kendall and Hench. He never even made it to the award ceremony in Stockholm that year.[16] While hunting in Uganda shortly before the prizes were to be bestowed, Hemingway was involved in not one but two plane crashes. The second mishap nearly killed him; he was left with injuries from which he never recovered: fractured skull; vertebral damage; dislocated shoulder; ruptures of the liver, spleen, and kidney; and serious head and arm burns.[17]

As with Philip Hench, the Nobel Prize seemed to aggravate a personality change in its recipient. Hench's change was subtle but definite. Hemingway's would be anything but subtle.

In 1950 John F. Kennedy, now the most famous patient ever to have Addison's disease, was, like Hemingway, in near-constant back pain. X-rays would eventually show that he was suffering from a fifth lumbar vertebra collapse. Publicly, the reason for his condition was highly speculative; sports injuries, his war wounds, and numerous politically correct diagnoses were offered publicly. In private, there was virtually no doubt what had happened. Kennedy's backbone had crumbled as a result of weakening from the DOCA pellets and daily oral doses of cortisone that he took to control his Addison's disease. Attempts to cut down or eliminate his steroids were proving futile. At the end of 1950, while on a trip to Japan, it's reported that he stopped taking his steroids and had an "Addisonian crisis"—a severe, acute worsening of his Addison's disease.[18] He would win his Senate race the next year, but all the political power in the world could not legislate away his painful spine.

In August 1954 Kennedy's back problems became critical; something aggressive had to be done. A team of Boston physicians examined him and recommended radical surgery. They suggested fusing the spine and sacro-iliac regions to add strength and stability to his back. Without surgery, it was possible Kennedy would eventually lose his ability to walk.

The bigger problem, however, was his cortisone use and Addison's disease. Steroids were threatening to cause two significant problems during surgery: acutely, the stress of surgery might trigger a worsening of his Addison's disease, which could quickly become fatal if additional steroids were not administered promptly. But Kennedy's daily use of steroids had left him with a suppressed immune system; he had a high risk of postoperative infection or impaired wound healing. These concerns explain why an operation of this magnitude had never been attempted on a patient with Addison's disease.

Kennedy accepted the operative risk, arguing to his friends that his painful existence wasn't worth living, and the risk was therefore justified. On October 21 a three-hour surgery was performed in Massachusetts, during which a metal plate was used to stabilize Kennedy's spine. One historian described it as "at best a limited success."[19] His hospital course was complicated by a severe urinary tract infection, which put him into a coma, causing a priest to once more administer last rites to Kennedy. He pulled through again, but not without more difficulties. As feared, the newly inserted plate in his back began to behave as if it were infected. Four months later, Kennedy underwent another spine operation, this time at New York Hospital, to remove the plate that had been so problematic to insert in the first place.[20]

Kennedy's personal physician would later remark how her patient "resented" the back surgeries, which had produced little, if any, relief, and "seemed to only make him worse." She recognized the role cortisone had played in producing Kennedy's ailments and tried to steer her patient clear of all steroids from then on. She failed. It's reported that Kennedy took testosterone to "keep up his weight," a practice that conceivably made JFK the first major politician to use steroids for purely anabolic purposes.[21]

Kennedy made the most of the time he spent recovering from his back surgeries—he wrote a book. Eventually published in 1956, *Profiles in Courage* became a best seller and established Kennedy, in the minds of many, as a "serious" writer. Like so many things associated with Kennedy, the book generated controversy. This time the brouhaha didn't involve the topic or material—*Profiles in Courage* was basically a retelling of the stories of several heroic U.S. senators—but rather the suspicion that the book had been "ghostwritten." Mike Wallace, famous for his later involvement with *60 Minutes* but then a young television journalist, went so far as to accuse Kennedy of using a ghostwriter. Clark Clifford, a lawyer hired by the Kennedy family, was able to persuade Wallace's bosses at the ABC network to do an on-air retraction. It mattered little. Over the ensuing years a consensus has developed among Kennedy historians that the book, while directed by Kennedy, was largely written by others.[22] Nonetheless, *Profiles in Courage* won the 1957 Pulitzer Prize for a biography.

If there had been no ghostwriting controversy swirling around *Profiles in Courage*, is it possible that JFK, like Hemingway, could have won his own Nobel Prize in Literature? After all, three years earlier another part-time writer and full-time politician, Winston Churchill, had won the Nobel in Literature for his similarly "idealistic" historical recountings. Was Kennedy perhaps more deserving of this recognition than the actual 1953 Nobel Prize winner, Albert Camus?

Absurd.

CHAPTER 29

Twilight

The fates are against you, Watson.

—SHERLOCK HOLMES, *THE REIGATE PUZZLE*

BURIED INSIDE THE JUNE 15, 1957, EDITION OF THE *ROCHESTER Post-Bulletin* was an article noting that "two long-time members of the staff of the Mayo Clinic, Drs. Philip S. Hench and Claude F. Dickson, have requested early retirement from the staff and have been granted the request by the Board of Governors, it was announced today." Hench was only sixty-one years old, still four years away from the mandatory retirement age that had sent Dr. Kendall to Princeton six years earlier. The article went on to state that Hench made the request because of the "steadily increasing commitments" he faced after winning the Nobel Prize.

What the local newspaper didn't mention was the possibility that Hench's position as a board member of the multimillion-dollar Kahler Corporation—his wife's family business—was conflicting with his work at the clinic. The article also politely ignored the most speculative rumor surrounding his early retirement from the clinic: the perception held by many that Philip Hench's personality was somehow deteriorating.

If Hench was drifting into a place of psychological uncertainty, the change was subtle; the great rheumatologist could still show flashes of his famous joviality. Dr. Emmerson Ward, a rheumatologist recruited to Hench's group in 1950 (and the chairman of the board of governors from 1964 to 1975), distinctly remembers a conversation with his boss shortly before Hench retired. According to Ward, Hench had a habit of showing up late to work and parking in a small lot near the clinic that was reserved for emeritus staff members. Catching him as he got out of his car one morning, Dr. Ward chided Hench that "Phil, this parking lot is for retired people, not members of the active staff." Without a moment's hesitation Hench replied, "Well, I'm not very active."[1]

Moments of mirth like this aside, Hench's increasing irritation with life in general was becoming more apparent since he'd won the Nobel Prize. Even his beloved forays into the history of yellow fever were no longer immune to outbursts of impulsive bad behavior. Hench had always been the polite, sincere, impeccably diplomatic sort, but he was now becoming short-tempered with anything annoying—and his list of annoyances was a long, growing one.

On February 24, 1954, Hench composed an elaborate multipage account of an unpleasant, ongoing series of petty interactions he'd had with the offspring of one of Walter Reed's associates. His rationale for launching this harangue was to document and justify "the generally uncooperative, suspicious or even antagonistic attitude" of the person in question and his family members; in doing so Hench unintentionally reveals his own increasingly strange behavior.[2]

Consider Hench's description of a particular occasion on which he apparently harassed his antagonist's widow. On this occasion Hench showed up unexpectedly at her home; he was hoping to examine a collection of Walter Reed paraphernalia in her attic—items to which her recently deceased husband had repeatedly denied him access. Unfortunately for Hench, on this occasion the woman refused to answer the door.

> I rang the bell for minutes at a time, pounded on various windows and called out to her through the door and different windows. . . . After 20 minutes of the above, I went to a nearby restaurant and phoned her. No reply. Then back to the house where I examined every door, including the side door and back basement door. They were all locked and she did not respond to my calls and knocks. Back to the restaurant; I phoned again and the line was busy. . . . I returned to the house for the third time, and then purposely let the outer screen door slam so as to make the decided impression of a definite final departure. I then walked down the street toward the streetcar stop but cut across the street and cut back, walking along a road in a deep ravine and then scrambled up a deep embankment, ending behind a service station, diagonally across the street from (her) home. I was hoping that if she was in the house and assumed that I had left, she might open the door to a mailman or some delivery man, where upon I would walk hastily across the street, but no luck. After waiting at the service

station for half an hour, I crossed the street to the restaurant and made two more phone calls. Once the line was busy, the second time there was no answer. For the fourth time, I went to the house and when there was no response to my knocking, I walked to the middle of the street in front of the house so that if she was peeping, she could see me taking a taxi.[3]

Why would Hench go to such embarrassing extremes to question a still-grieving widow about some belongings her late husband's father might have possessed—and clearly did not want Hench to examine? More important, why couldn't the impeccably proper doctor see how odd and inappropriate his own actions were becoming?

Hench's friends believed that his irritability and irrational behavior were somehow related to the criticisms being raised about cortisone. Hench never claimed that cortisone was a perfect drug; he readily recognized its potential for harm, its inability to cure late-stage rheumatic disease (or indeed, to "cure" anything—it only reduced inflammation), and all the other uncertainties inherent in its use. But despite his repeated and emphatic public disclaimers, the fact that he had won the Nobel Prize meant (to many members of the public and the scientific community) that he was an advocate for the drug. Hench began to believe that anytime cortisone was criticized, fairly or not, the criticism was a personal attack against him.[4] The unbridled enthusiasm that had greeted the initial announcements regarding cortisone was now tempered by increasingly critical findings and comments. And they were coming from all directions—friends and foes alike.

Initially, the anticortisone sentiments were based on anecdotes. Then data from clinical trials began to appear. Most studies, including ones performed outside of Mayo Clinic, confirmed the efficacy of cortisone in treating rheumatoid disease, but the newest of these demonstrated a higher incidence of adverse side effects than previously suspected. A testy editorial in the *Journal of the American Medical Association* from February 4, 1950, pointed out that "it is not uncommon when powerful new agents are discovered, the immediate value of these substances is overemphasized in the popular press."[5]

Published concerns quickly mounted. Reports of cortisone-induced psychotic states,[6] as well as diabetes, began appearing in the literature with regularity.[7] Cortisone was depicted as a "two-edged sword" whose use "demanded discrimination."[8] The *New England Journal of Medicine* went so far as to argue that the current scarcity of cortisone "might actually be a blessing in disguise" because of cortisone's potential to cause more harm than good when inexperienced doctors prescribed large doses for their patients.[9]

Even Hench's closest friends occasionally critiqued the use of the drug. Dr. Richard Freyberg, one of the original physicians to meet at Hench's house and conduct the earliest trials with cortisone, was now warning the public that "indefinitely prolonged use of cortisone . . . is not a practical solution."[10] This criticism, however, seems gentle in comparison to that leveled by Russell Cecil, who frequently referred to cortisone as "glorified aspirin" and noted to Hench's chagrin that "hyperadrenalism is not the answer to the rheumatoid arthritis problem. . . . Hench and Kendall have only given us two more drugs to fumble with."[11]

Then an academic A-bomb hit cortisone—and Hench. In 1954 a British multicenter study was published; sixty-one patients with rheumatoid arthritis had been placed in a crossover trial in which cortisone was tested against aspirin.[12] To everyone's surprise, the results failed to show a significant difference between the two groups. Although serious concerns would later be raised regarding the size, design, and interpretation of this trial, the damage was done—and it was front-page news. Philip Hench was livid; not only did he disbelieve these conclusions, but the publication based on these findings was written by authors whom he had "numbered among his greatest friends."[13] Hench began to refer to some of these people as "traitors" and declined any further association with them, leaving many of his former colleagues bewildered by his snubs.

The emotional toll on Hench was considerable. Dr. Emmerson Ward commented that "Phil seemed to go through a depression. Why? Because cortisone had side effects. This wasn't Phil's fault. Everybody knew that. But Phil felt like everybody thought it was his fault."[14]

Dr. John Glyn of London was a longtime friend of Phil Hench.[15] Although Glyn's writings demonstrate his sincere belief that "Philip Showalter Hench was the most remarkable man I have ever met," he occasionally describes his friend as a man embittered by the imperfections inherent to cortisone:

> Despite the cautious and modest attitude which (Hench) originally adopted toward his discovery . . . he later seemed to find it extremely difficult to accept that cortisone was not the ultimate solution to the treatment of rheumatoid arthritis. He argued that the side effects of cortisone therapy were not "inevitable," indeed that they were the result of failing to tailor the dose accurately to

the patient's requirements. Toward the end of his life he fell out not only with his British friends but also with many American colleagues. Howard Polley and Charles Slocomb, who were his close collaborators to the end of his life, were loyal but eventually concluded that he suffered a profound personality change in the 1950's. . . . He seems in fact to have tragically become a victim of his own fame and he certainly achieved less happiness from his brilliant labors than he deserved.[16]

The results of the British cortisone trial were especially problematic for Hench, a self-admitted major-league Anglophile. He had recently been invited to his beloved England to address a meeting of the British Medical Association (BMA); this should have been a well-received invitation triggering an immediate acceptance by Hench. Instead, a low-grade creeping semiparanoia was getting the best of him. Shortly after the unfavorable British cortisone trial results were published, Glyn describes receiving a most unusual letter from Hench: "A few weeks later, there was a totally unexpected repercussion (of the trial results) in the form of a letter to me from Philip Hench, asking me whether I thought the atmosphere in Britain was propitious for him to accept an invitation to come and address a BMA meeting. He was not prepared to come if there was any risk of being heckled."[17]

Heckled? Hench's fear was, of course, unfounded. Fortunately, he overcame his irrational concern, made the trip to England, gave the talks (which were well received), and eventually began to visit the other side of the Atlantic on a regular basis. In time he seemed to prefer Europe to the United States. His son confirmed this impression and offered his own speculation on the reasons for his father's preference for the Continent.

> Having been with him and Mom on most of his summertime European trips in the early 1960s, I am convinced he enjoyed the status and adulation he could find abroad, because he no longer had as much of it at home. A lot of other famous and accomplished people also have found themselves better understood and happier anywhere but home. In Europe especially, he was free to exercise his more autocratic tendencies. He was what we would today call a "control freak." He was also charismatic. Like other people I have known, he easily became the center of attention in any crowd—and didn't much care for it if someone else challenged the spotlight.[18]

The revolution in Cuba seemed to dampen Hench's interest in the yellow fever story, but his increasing fascination with Europe revived his long-standing enthusiasm for Sherlock Holmes. Hench was a member of the "Norwegian Explorers," a Holmes appreciation group formed in Minnesota in 1948.[19] While visiting Switzerland with his wife in 1953, Hench was shocked to learn that the residents of the city of Meiringen—the location of Reichenbach Falls—had no idea of the literary significance of their home. Hench, with help from the Sherlock Holmes Society of London, erected a plaque near the falls in 1957; it commemorated the famous battle between Holmes and Professor Moriarty that had taken place there.[20]

It appears there were still a few ungerminated seeds of joy left in Philip Hench.

But the situation was becoming uglier for another recent Nobel Prize winner. Ernest Hemingway's existence, far more than Philip Hench's, was turning dark. For Hench, depression was entering early twilight. In contrast, Hemingway was at the midnight of mental illness.

Things had been going downhill rapidly for Hemingway. When Castro came to power in 1959, Hemingway opted to remain in Cuba, having no reason to fear the regime change. Although his relationship with Castro's government remained cordial, living conditions in Cuba began to deteriorate. Hemingway ultimately abandoned Finca Vigía, perhaps the only place he had ever felt truly comfortable, and returned to the United States. His overtly friendly relations with Castro did little to endear him to the U.S. government; worse, he became a "person of interest" to FBI director J. Edgar Hoover.[21]

Whether and to what extent Philip Hench became depressed later in life is debatable. Not so for Hemingway. His depression was becoming increasingly severe. The great writer began drinking heavily—even by his own formidable standards—and started talking openly of committing suicide. Mary, his wife, took his threats seriously. Determined to obtain help for her husband, she arranged for him to undergo psychiatric evaluation.

In November 1960 John F. Kennedy—now, more than ever, the world's most famous person with Addison's disease—was elected president of the United States. He invited Ernest Hemingway to his January 1961 inauguration in Washington, D.C. But the Nobel Prize–winning, world-famous author was unable to attend.[22] Why? Because in January 1961 Hemingway was locked in a hospital psychiatric unit—at the Mayo Clinic.[23]

The 1960s were not kind to Philip Hench. His emotional edginess was now accompanied by blatant physical deterioration. His friend John Glyn concedes that "certainly he lost a huge amount of weight and he also developed severe diabetes."[24] Luckily, diabetes was treatable; after all, a Nobel Prize had been awarded because of the hormone—insulin—that could control the dreaded disease. As a Nobel Prize winner himself—and with the distinction of having helped to discover a powerful, therapeutic hormone of his own, Hench should have intellectually embraced insulin and its life-saving therapeutic properties.

Except he didn't. For reasons that are unclear, Philip Hench wouldn't take insulin himself.[25] His son John later commented, "As for the question about not taking insulin, however, I do recall that he was not a particularly good patient—probably not the only MD to exhibit such behavior."[26]

Hench's untimely deterioration was steady—and ultimately fatal. Philip Hench died on March 30, 1965, while vacationing in Ocho Rios, Jamaica. He was sixty-nine years old. His Rochester obituary cited pneumonia as the cause of death, and references to his diabetes were generally purged.[27] Hench never completed the long-awaited book on Walter Reed and yellow fever that he had researched for more than three decades. His wife, Mary, eventually gave his extensive collection of material to the University of Virginia—Walter Reed's alma mater and the institution at which Hench's brother, Atcheson, was an English professor. The Henches' huge and extremely valuable collection of Sherlock Holmes material was ultimately bequeathed to the University of Minnesota.

Remembering the good with the bad, John Glyn summed up the life of the famous rheumatologist in this way:

> Philip Hench had a charismatic and generous personality. He was a man of diverse and enthusiastic interests outside medicine. His sensitivity on the subject of his seminal contribution to medicine was unfortunate, and it undoubtedly marred the pleasure he should have derived from his fame and from the Nobel Prize for Medicine in 1950. It was especially unfortunate in view of his original intention to present his discovery as an investigative tool rather than as a therapeutic breakthrough.
>
> The clinical usefulness of cortisone in rheumatology remains controversial 50 years after the event, but without doubt its discovery transformed the specialty from its Cinderella status of the B.C. (before cortisone) era. Its significance in general medicine remains beyond dispute.[28]

CHAPTER 30

The End of the Show

Come, friend Watson, the curtain rings up for the last act.

—SHERLOCK HOLMES,
THE ADVENTURE OF THE SECOND STAIN

ERNEST HEMINGWAY WASN'T OUT OF THE HOSPITAL FOR LONG. Depression still squeezed him in its serrated jaws. In the spring of 1961, just as Percy Julian was selling his steroid-manufacturing company, Julian Laboratories, to Smith, Kline, and French for more than $2 million,[1] Hemingway attempted suicide again and wound up rehospitalized in Saint Marys Hospital. His subsequent treatment reportedly involved more electroconvulsive therapy.[2]

Hemingway's seemingly premature release from this hospitalization has provided endless fodder for conspiracy buffs. Why did his doctors let him out when they did? Had he feigned just enough psychiatric improvement to get himself dismissed?[3] Did he leave against medical advice? Or was his depression actually under good control at the time of his discharge, only to relapse later as depression often does?[4]

Once freed from Saint Marys, the writer departed Minnesota for the last time and withdrew to his retirement home in Ketchum, Idaho. Hemingway, like Lew Sarett—the man who first synthesized cortisone—retired in Idaho to pursue hunting and fishing. But there would be no more of either for the Nobel-winning writer. On July 2, 1961,[5] Ernest Hemingway unsheathed his favorite shotgun, a 12-gauge model purchased from Abercrombie & Fitch, and killed himself.[6]

Ortho Pharmaceutical Company marketed its first birth control pill in 1963: 2.3 million women were already using the increasingly popular prescription steroid contraceptives. Nobody remembers where they were on the day that the new pill was introduced.

But everybody over fifty-five years of age remembers where they were on November 22, 1963, at 12:30 P.M. central standard time—the moment John F. Kennedy was shot. More precisely, shot twice. According to the Warren Commission, the first shot, which passed through Kennedy's neck, might not have been fatal. But rather than collapsing into the relative protection of his limousine's deep rear seat, Kennedy remained an erect target for the next round. Moments later a second bullet shattered the right side of his skull.

Authorities have speculated that Kennedy remained upright and exposed to the second round because of the rigid back brace he wore to support a spine that had been ravaged by chronic steroid use.[7] If so, Kennedy's Addison's disease—and the cortisone that may have both caused and treated it—played a secondary role in his death. Another small controversy in the most controversial assassination of the century.

Luis Alvarez, the problem child from Rochester, Minnesota—now a nationally famous scientist—was soon brought in as a part of the JFK investigation. No stranger to presidents (he had developed an indoor golf training machine to help improve the game of President Eisenhower), Alvarez was invited by the Warren Commission to provide input on Kennedy's assassination. As one of the nation's leading particle physicists, Alvarez was considered an expert in the areas of "trajectory" and "ballistics." His assignment from the committee was to analyze the famous Abraham Zapruder film showing Kennedy's head at the moment of projectile impact. Alvarez's testimony supported one of the most critical conclusions of the Warren Commission—the opinion that the fatal shots had indeed come from behind Kennedy despite the sharp, backward movement of the president's head at the time of impact.[8]

Years later, in the September 1976 issue of the *American Journal of Physics*, Alvarez published a detailed article describing "the odd behavior" of Kennedy's head.[9] Using experiments performed on melons, as well as his own simple theories regarding the "law of conservation of momentum," he demonstrated that the paradoxical backward motion of Kennedy's head was the result of recoil caused by the explosive, forward ejection of the president's brain as a result of the bullet's impact. It was a careful, detailed, hard-to-refute explanation for one of the most controversial pieces of expert testimony provided to the Warren Commission.

San Francisco was known for its "Summer of Love" in 1967. Now it was time for some other Bay Area "locals"—with their own steroid connections—to grab the headlines.

Dr. Norman Shumway of nearby Stanford University performed the first successful adult heart transplant in the United States on January 6, 1968; the achievement was made possible by prednisone, a newly available longer-acting version of cortisone with the power to fight organ rejection. Bill Toomey, a thirty-six-year-old with a graduate degree from Stanford, won the 1968 Olympic decathlon in Mexico City; it's widely quoted that one out of three American athletes at the games used performance-enhancing steroids.[10] And why not? As Toomey himself pointed out, performance-enhancing steroids weren't banned from the Olympics until 1972.[11]

But the best was yet to come. Later that year Luis Alvarez—now a resident of Berkeley, California—received a rare, solo Nobel Prize for his work with the hydrogen bubble chamber and for his use of this device to discover and characterize numerous subatomic particles.[12] Little Luis Alvarez, once the teenage terror of Rochester, Minnesota, was no longer merely one of the most famous physicists in the United States. Overnight, he'd become one of the most famous physicists in the world.

Incredibly, Luis Alvarez entered the international spotlight one more time in 1980 for collaborative work undertaken with his son Walter (named, of course, after his famous Mayo Clinic grandfather), a geologist also on the faculty of the University of California at Berkeley. The father-son team proposed a theory to explain the sudden extinction of the dinosaurs sixty-five million years earlier.[13] This theory postulated that a meteor, or some other large extraterrestrial mass, struck the earth and created a global climate disturbance that disrupted the balance of life. The physicist and the geologist combined their talents to support this theory by demonstrating high concentrations of iridium (a substance commonly found in extraterrestrial matter) in rock layers dating to the period in question.[14] Luis and Walter's hypothesis remains, to this day, the most widely accepted explanation for the abrupt end to the Age of Reptiles. The possibility that another Nobel Prize might someday be awarded for this work, while remote, is not out of the question.

Luis Alvarez—son of the famous physician Walter Alvarez, Mayo Clinic machine-shop summer help, radar expert, atomic bomb builder, and Nobel Prize laureate—died of cancer in 1988.

Edward "Nick" Kendall was feeling old, sad, and worn out during the weeks before his death. The past two decades at Princeton had given him peace of mind—and a chance to complete his memoirs (*Cortisone: Memoirs of a Hormone Hunter*)—but the time had not yielded any more scientific break-throughs. Kendall never made an important chemistry discovery during his tenure in New Jersey, but at least he got the chance to keep working at something that excited him. For a hopeless workaholic like Kendall, this opportunity was its own reward. After sixty-two years of active research (he had earned his PhD in 1910 and had been in the lab nonstop ever since), "The Chief" was physically and emotionally drained.

The eighty-six-year-old Kendall was still working as a private consultant to the pharmaceutical industry. On May 1, 1972, while on an official visit to Merck & Company—his original partner in the quest for cortin twenty years earlier—he arranged to have lunch with several of his old friends. At one point during the informal noon meal he went to the blackboard to write a formula and was suddenly stricken with chest pain.[15] The chemist died three days later of "coronary failure." Rebecca Kendall, by all accounts a charming woman who had battled depression for most of her life, was understandably lost without her partner of the last six decades. She died nine months later.

One of Kendall's closest Mayo associates, Dwight Ingle, commented bluntly about the mentor he both admired and respected.

> I have argued that a man should be remembered for his best personal qualities and for his achievements rather than for his foibles and failures, but I cannot write of Dr. Kendall without describing his weaknesses as well as his strengths. To do so would create an image of a person who never existed. His greatness lay in his ability to select important goals that were achievable, to persevere toward them during periods of adversity and disappointment, and to select gifted associates.[16]

Ingle's words seem to be a fitting, concise, and ultimately accurate summary of the great chemist.

By 1977 the Food and Drug Administration (FDA) had confirmed that estrogen raises a woman's risk of cancer and blood clots. The agency was preparing a comprehensive warning that all pharmacists would be required to provide to patients at the time of drug disbursement. But Dr. Walter Alvarez didn't live to see this mandate implemented. Philip Hench's close friend, adviser, and confidant—and now the father of a Nobel Prize winner—died that year.

Walter Alvarez had published his memoirs, *Incurable Physician*, in 1963. This book, like almost everything else he wrote, became a big seller. Of course it did; by the time of his death Walter Alvarez had become one of the most famous Mayo Clinic doctors since the Mayo brothers themselves.[17]

Russell Marker, the most unpredictable and colorful figure in the history of cortisone—perhaps in the whole history of chemistry?—eventually emerged from his self-imposed social exile. The man who seemingly trusted no one, especially when money was involved, became a philanthropist; in 1984 he even endowed a lectureship in chemistry at Penn State that bears his name. Russell Marker died in 1995 at the age of ninety-five, and by all accounts he lived out his final decades as a happy, crazy old man.

Not a bad ending for someone who almost became a "urine boiler."

Notes

INTRODUCTION

1. H. Clapesattle, *The Doctors Mayo* (New York: Pocket Books, 1954), 144–64.
2. NOAA's National Weather Service Weather Forecast Office, La Crosse, Wis., "The Rochester, MN Tornado of 1883" (Washington, D.C.: National Oceanic and Atmospheric Administration), updated November 8, 2005, http://www.crh.noaa.gov/arx/events/rst_tor1883.php.
3. Mayo Clinic, Rochester, Minn., "Tornado Strikes Rochester—Saint Marys Hospital Opens: Born in a Storm," Mayo Foundation for Medical Education and Research, 2001–2007, http://www.mayoclinic.org/tradition-heritage/born-storm.html.
4. Violated and unforgiving, Cascade Creek waited over 100 years to avenge this meteorological humiliation. Unfortunately, the final reckoning came when I was training as a medical resident in Rochester. Back in those days my wife and I owned a small gray cedar shake house with a backyard abutting the infamous creek, which by then was little more than a slow-flowing drainage ditch. Our home was the gem of the neighborhood, an architectural classic located just a few doors away from the house in which Dr. Benjamin Spock, the famous baby expert, had lived during his short tenure on the faculty of the Mayo Clinic half a century earlier. One fall evening, as we slept peacefully, Cascade Creek took advantage of unseasonably heavy rains to once again leave its banks. But this time there would be no whirling around in the breeze or flinging of fish. On this night the "Nile of Rochester" crept silently out of its bed and flooded our low-lying abode. As the house slowly filled with black water and dirt once bloodied by the victims of the tornado, my connection to the tragic event of 1883 was cemented.
5. Mayo Clinic, Rochester, Minn., "Tornado Strikes Rochester—Saint Marys Hospital Opens: Born in a Storm."
6. Mayo Clinic, Rochester, Minn., "Tornado Strikes Rochester—Saint Marys Hospital Opens: Winds of Change: An Improvised Hospital," Mayo Foundation for Medical Education and Research, 2001–2007, http://www.mayoclinic.org/tradition-heritage/winds-change.html.

7. Clapesattle, *Doctors Mayo*, 146.
8. Ibid., 148.
9. Mayo Clinic, Rochester, Minn., "Tornado Strikes Rochester—Saint Marys Hospital Opens: Winds of Change."
10. G. Rienzi, "Celebrating a Dual Legacy of a Johns Hopkins Luminary," *JHU Gazette* 34, no. 9 (October 25, 2004).
11. Mayo Clinic, Rochester, Minn., "Tornado Strikes Rochester—Saint Marys Hospital Opens: Partnering with the Sisters of Saint Francis," Mayo Foundation for Medical Education and Research, 2001-2007, http://www.mayoclinic.org/tradition-heritage/partnering-sisters.html.
12. Mayo Clinic, Rochester, Minn., "Tornado Strikes Rochester—Saint Marys Hospital Opens: The New Saint Marys Hospital," Mayo Foundation for Medical Education and Research, 2001–2007, http://www.mayoclinic.org/tradition-heritage/new-saint-marys.html
13. Clapesattle, *Doctors Mayo*, 118, 122.
14. M. W. King, "The Medical Biochemistry Page" (Indianapolis: Indiana University School of Medicine, 1996–2009), updated January 20, 2009, http://www.themedicalbiochemistrypage.org/steroid-hormones.html.
15. History Channel Web Site, "Hormone," A&E Television Networks, 1996–2008, http://www.history.com/encyclopedia.do?articleID=212147.
16. S. Whitehead, "Editorial," *Endocrinologist* 75 (Spring 2005), http://www.endocrinology.org/sfe/endocrinologist/end07502.pdf.
17. A. Smith, "Lipitor: Cause for Concern," CNNmoney.com, January 19, 2006, http://money.cnn.com/2006/01/19/news/companies/Pfizer/index.htm.
18. J. M. Schrof, "Pumped Up," *U.S. News & World Report*, May 24, 1992, http://www.usnews.com/culture/articles/920601/archives_017805.htm.
19. "Cortisone: The Limits of a 'Miracle,'" *Nutrition Health Review* (September 22, 1991), http://www.encyclopedia.com/doc/1G1–11612074.html.
20. For example, a bigger dose of the same natural testosterone that gives Barry White his deep bedroom voice is another man's one-way ticket to Schwarzeneggerville.
21. A. Smith, *The Wealth of Nations* (New York: Modern Library, 1937).
22. Thus the rationale for having each chapter begin with a scene-setting quote by Sherlock Holmes—Hench turns out to be one of the world's leading authorities on the fabled detective. Other characters and events, some of whom may (initially) seem, like Holmes, to be tangential to the cortisone story are similarly included for the relevant perspective, context, or background they provide.

CHAPTER 1. ADDISON AND HIS DISEASE

1. C. D. Wehner, "Curriculum Vitae: The Real Addison's Disease," 2001, http://wehner.org/addison/CV/index.htm.
2. On June 29, 1861—a year to the day after the death of Thomas Addison—a seminal event in the history of cortisone took place in Le Sueur, Minnesota. William James Mayo, the son of Dr. William Worrall Mayo and the same young doctor who was nearly killed along with his brother in the great Rochester tornado, was born.
3. Haggis, a traditional Scottish casserole, is prepared by stuffing assorted sheep's organs back into the animal's stomach and boiling them until they're suitably wretched; it's possibly the most vile entrée on the menu in a country where the competition for "most vile entrée" is pretty much a national sport.
4. Wehner, "Curriculum Vitae."
5. Ibid.
6. Ibid.
7. Ibid.
8. Perhaps Addison's parents ran a flower-selling, rather than a flour-selling, business? O. D. Enersen, "Who Named It? Thomas Addison," 1994–2009, http://www.whonamedit.com/doctor.cfm/68.html.
9. National Information Centre for Metabolic Diseases, CLIMB (Children Living with Inherited Metabolic Diseases), "CLIMB Update: So, What Is Addison's Disease?" 2, no. 7 (March 2004), http://www.climb.org.uk/Magazines/addison.doc.
10. Pearce, "Thomas Addison," 297–300. Sadly, writing shorthand notes to yourself in a dead language was considered just as strange back then as it would be today, and rather than impress his peers, it reinforced their notion that Addison was an 1812 version of the pencil-necked little geek. This reputation took its expected toll on his social life—and probably explains why Addison didn't marry until he was over fifty years old.
11. National Information Centre for Metabolic Diseases, CLIMB (Children Living with Inherited Metabolic Diseases), "CLIMB Update." Forty-four years later a budding author and fellow Scottish countryman, Arthur Conan Doyle, would likewise graduate from this same medical school.
12. The Greeks needed only four elements—earth, wind, fire, and water—to explain the composition of the entire universe. Sharing certain characteristics with these four elements were specific "humors" (fluids) of the body—black bile, yellow bile, blood, and phlegm. On the basis of these four humors, the Greeks believed the body's construction (anatomy) and function (physiology) could be understood. The "humoral theory" explained disease as the

result of humoral imbalance—when the humors were in harmony, health ensued. When they weren't, you got sick. This paradigm provided a rationale for medical therapy—to cure disease, you needed to restore humoral balance. But how? The Greeks proposed two methods. First, by "direct restoration" of humoral balance through bloodletting, starving, purging, and so forth. Second, by "indirect restoration" using medicines to augment or oppose humors as needed.

As science and knowledge evolved over the next 3,000 years, the four humors turned into proteins, fats, carbohydrates, DNA, and all of the other substances we now know constitute our bodies. Some of these biological building blocks were known in Addison's time, and by the 1800s it was understood that these elements, not the four classic humors, needed balance to produce health. Claude Bernard (1813–1878), the great French physiologist, summed this up with the concept of the milieu interior, or the idea that an internal environment surrounds each cell in the body, and that this environment must be rigorously regulated with regard to oxygen, salt, waste buildup, and so forth. Alter the intercellular conditions too much and the cell—followed by the rest of the organism—dies.

13. Wehner, "Curriculum Vitae."
14. Ibid.; Thomas Addison Unit of St. George's Hospital, "Historical Background," http://www.addison.ac.uk/historical/historical.htm.
15. K. Løvås and E. Husebye, "Addison's Disease," *Lancet* 365, no. 9476 (June 11–17, 2005): 2058–61.
16. National Information Centre for Metabolic Diseases, CLIMB (Children Living with Inherited Metabolic Diseases), "CLIMB Update."
17. Enersen, "Who Named It? Thomas Addison."
18. Sherlock Holmes was, of course, the literary creation of the aforementioned Edinburgh medical graduate, Arthur Conan Doyle.
19. Wehner, "Curriculum Vitae."
20. Addison's words were moving: "A considerable breakdown in my health has scared me from the anxieties, responsibilities, and excitement of the profession; whether temporarily or permanently cannot yet be determined; but, whatever may be the issue, be assured that nothing was better calculated to soothe me than the kind interest manifested by the pupils of Guy's Hospital during the many trying years devoted to that institution." Wehner, "Curriculum Vitae."
21. Pearce, "Thomas Addison," 297–300.
22. C. D. Wehner, "The Real Addison's Disease: Constitutional and Local Effects of Disease of the Supra-Renal Capsules by Thomas Addison, MD," Charles Douglas Wehner, 2001, http://wehner.org/addison/x4.htm.

23. J. R. Hiatt, "The Conquest of Addison's Disease," *American Journal of Surgery* 174, no. 3 (September 1997): 280–83.

24. Hardin Library for the Health Sciences, "Addison's Disease Pictures/Thomas Addison," University of Iowa, 2009, updated November 13, 2008, http://www.lib.uiowa.edu/hardin/md/ui/addisons/index.html.

25. Wehner, "Real Addison's Disease."

26. W. Jeffcoate, "Thomas Addison: One of the Three 'Giants' of Guy's Hospital," *Lancet* 365, no. 9476 (June 11–17, 2005): 1989–90.

27. AIM25: Archives in London and the M25 Area, "Cope, Sir Vincent Zachary (1881–1974)," AIM25, 1998–2008, 2008, http://www.aim25.ac.uk/cgi-bin/search2?coll_id=3291&inst_id=8.

28. "Addison's Disease and Two of Its Famous Sufferers," *Ockham's Razor with Robyn Williams*, radio program, Australian Broadcasting Corporation, aired March 1, 1999.

29. *Jane Austen's Illness*, http://www.orchard-gate.com/bmj.htm.

30. Ibid.

31. Ibid.

32. Ibid.

33. For Brown-Séquard, this landmark study was a temporary high point in a career that would fluctuate like the price of gas. He was elected to the Royal Society of London in 1861—another high point—but feeling the pressure of professional practice, he left for Paris, and later moved to Philadelphia—a definite low point. His difficulty with the English language was an albatross for a man who earned his living speaking and giving lectures. During a four-month tenure on the faculty at the Medical College of Virginia—another low point—his talks were described as "not very unlike an attack of spasmodic asthma," causing one historian to note that "the agony of trying to make himself understood was, if anything, topped by the agony of his listeners trying to comprehend." After acquiring competence in English, he eventually became the chairman of the Physiology and Pathology Department at Harvard Medical School, a position he occupied from 1864 to 1867. O. D. Enersen, "Who Named It? Charles-Édouard Brown-Séquard," 1994–2009, http://www.whonamedit.com/doctor.cfm/977.htm.

34. Brown-Séquard's involvement with this field does not end here. Despite relatively draconian leanings on numerous social issues (he once told a Harvard Medical School class that laboratory animals didn't mind being vivisected), Brown-Séquard was an unabashed liberal when it came to the pursuit of science. In the tradition of another famous (future) member of the Harvard faculty, Dr. Timothy Leary, Brown-Séquard was willing to use himself as an experimental guinea pig—at least when the research involved a combination

of sex and drugs. His work in this area would eventually lead Brown-Séquard to reach a bizarre conclusion regarding an entirely different set of steroid-producing glands from those associated with Addison's disease. Enersen, "Who Named It? Charles-Édouard Brown-Séquard."

CHAPTER 2. INTRODUCING DR. KENDALL

1. E. C. Kendall, *Cortisone: Memoirs of a Hormone Hunter* (Edward C. Kendall, 1971), 31–32.
2. D. J. Ingle, *Edward C. Kendall, 1886–1972: A Biographical Memoir* (Washington, D.C.: National Academy of Sciences of the United States, 1975), 253.
3. Kendall, *Cortisone*, 1–7.
4. Ibid., 4.
5. Ibid., 10.
6. Perhaps because his social résumé was somewhat thin, Kendall was especially proud of one particular episode of uncharacteristically rowdy behavior. It was the night he and three fraternity brothers gained access to the Manhattan Bridge (which was still under construction and not yet open to traffic) and crossed the East River on its swaying catwalk. The catwalk, built of greasy two-by-fours strung together by long ropes and suspended beneath the ironwork of the unfinished bridge, was used by the bridge workers to haul construction materials. It was slippery and dangerous by day, and even more so by night. Being one of the first "civilians" to cross the river via this route may have been the most—perhaps only?—edgy or dangerous thing Kendall ever did in his life, and he was clearly proud of it.
7. M. A. Shampo and R. A. Kyle, "Edward C. Kendall: Nobel Laureate," *Mayo Clinic Proceedings* 76 (2001): 1188.
8. Kendall, *Cortisone*, 15.
9. Typical of those who know little about the practice of medicine, the administrator wondered "how the physicians in the hospital could be expected to treat patients unless they knew what they were giving them for breakfast." Kendall, *Cortisone*, 25.
10. Ibid.
11. Ibid., 25–26.
12. Had the lovebirds opted to go to the movies rather than take a trip down the aisle, they likely would have seen Cecil B. DeMille's *Carmen*, staring Kendall's favorite opera singer, Geraldine Farrar, in the title role. The movie was commercially successful (it made more at the box office than Theda Bara's visually

sultry competing version of *Carmen*, released that same year), which was all the more amazing because people were actually paying to see an opera with no singing. "HERO: The Official Gateway to Higher Education and Research in the UK," Newcastle-upon-Tyne: Hero Ltd. Lights, camera, Carmen, http://www.hero.ac.uk/uk/research/archives/2002/lights_camera_carmen 1664.cfm. "Talkies" were still ten years away.

13. Nobelprize.org. "The Nobel Prize in Physiology or Medicine 1909 for His Work on the Physiology, Pathology and Surgery of the Thyroid Gland: Emil Theodor Kocher," Nobel Web AB, 2009, http://www.nobelprize.org/nobel_prizes/medicine/laureates/1909/index.html.

14. O. H. Clark and P. Huarte, "Ambulatory Thyroid Surgery: Unnecessary and Dangerous," *Journal of Clinical Endocrinology and Metabolism* 83, no. 4 (1998): 1100–103.

15. Ibid.

16. Nobelprize.org. "Hamberger B. Emil Theodor Kocher," Nobel Web AB, 2009, http://nobelprize.org/nobel_prizes/medicine/articles/kocher/index.html.

17. Clark and Huarte, "Ambulatory Thyroid Surgery."

18. Ibid.

19. T. M. Habermann, R. E. Ziemer, and J. C. Lantz,"Images and Reflections from Mayo Clinic Heritage," *Mayo Clinic Proceedings* 77 (2002): 1182.

20. "Doctor Charlie," *Time*, June 5, 1939.

21. Clark and Huarte, "Ambulatory Thyroid Surgery."

22. Ingle, *Edward C. Kendall*, 249.

23. Ibid., 254.

24. Ibid., 256.

25. Ibid., 255.

26. Ibid., 264.

27. Ibid., 253.

28. Ibid., 263.

29. Ibid., 264.

30. Ibid., 254.

31. Sherlock Holmes once observed that "genius is an infinite capacity for taking pains." Ingle would have likely considered this to be a perfect summary of Kendall's brand of genius.

32. Ingle, *Edward C. Kendall*, 253–54.

33. R. D. Simoni, R. L. Hill, and M. Vaughan, "The Isolation of Thyroxine and Cortisone: The Work of Edward C. Kendall," *Journal of Biological Chemistry* 277 (May 2002): e10.

34. With the puzzle of the thyroid solved, the glands of the body, including the ovaries, testicles, adrenal medulla, pituitary gland, and pancreas, were now

largely understood. Only one endocrine structure still posed a major physiological mystery. Kendall would eventually solve the riddle of the adrenal cortex, but his journey would not start until he was visited by a Hungarian chemist—and future Nobel Prize winner—a few years later.

CHAPTER 3. LIFE AFTER THE THYROID

1. S. R. Ginn and J. A. Vilensky, "Experimental Confirmation by Sir Victor Horsley of the Relationship Between Thyroid Gland Dysfunction and Myxedema," *Thyroid* 16, no. 8 (August 2006) 743–47; R. C. Hamdy, "The Thyroid Gland: A Brief Historical Perspective," *Southern Medical Journal* 95, no. 5 (May 2002): 471–73.

2. The American Physiological Society: Integrating the Life Sciences from Molecule to Organism, "Timeline of Physiology—Endocrinology and Metabolism," (Bethesda, Md.: American Physiological Society, 2008), http://www.the-aps.org/press/endotime/index.htm.

3. Ibid.

4. Ibid.; M. J. Waters, H. N. Hoang, D. P. Fairlie, R. A. Pelekanos, and R. J. Brown, "New Insights into Growth Hormone Action," *Journal of Molecular Endocrinology* 36, no. 1 (February 2006): 1–7.

5. Essortment.com: Information and Advice You Want to Know, "Banting and Best Biography" (Round Rock, Tex.: essortment.com, 2005), http://www.essortment.com/bantingbestdia_rnoc.htm.

6. "The Board of Wisdom," http://www.boardofwisdom.com/mailquote.asp?msgid=4145.

7. PBS.org, "A Science Odyssey: People and Discoveries: Banting and Best Isolate Insulin" (Boston: WGBH, 1998), http://www.pbs.org/wgbh/aso/databank/entries/dm22in.html; Nobelprize.org, "Frederick G. Banting: The Nobel Prize in Physiology or Medicine 1923," Nobel Web AB, 2009, http://nobelprize.org/nobel_prizes/medicine/laureates/1923/banting-bio.html.

8. History wired, "Air Conditioning," Washington, D.C.: Smithsonian Institution, http://historywired.si.edu/detail.cfm?ID=189. At least Canada, unlike the United States, did not have Prohibition and thus a stein of freshly brewed, cold, sudsy, lifesaving liquid refreshment was always available.

9. Essortment.com: Information and Advice You Want to Know, "Banting and Best Biography."

10. "The Discovery of Insulin: A Canadian Medical Miracle of the 20th Century," http://www.discoveryofinsulin.com/Home.htm.

11. Mount Allison University Centre for Canadian Studies, "A Great Canadian Breakthrough: The Discovery of Insulin" (Sackville, N.B.: Centre for Canadian Studies at Mount Allison University, 2001), http://www.mta.ca/faculty/ arts/canadian_studies/english/about/ study_guide/doctors/insulin.html.

12. Despite their success, the team studying insulin squabbled nonstop. "Paranoia and insecurity ran rampant in the lab." CBC Digital Archives, "Egos and Ownership: Did You Know?" (Toronto: Canadian Broadcasting Corporation, 2009), http://archives.cbc.ca/IDC-1–75–702–4061/science _technology/diabetes/clip2). Banting and Best lived in constant fear that their relative inexperience would be taken advantage of by Collip and Macleod, both of whom were skilled senior scientists and ruthless political animals. Battles involving egos, distrust, jealousy, and ungratefulness erupted. Banting became depressed, causing him to temporarily terminate the relationship with his fiancée. Banting eventually married this woman—Marian Robertson—in 1924. They divorced in 1932. But failure never thwarts a great scientist. Banting tried for marital bliss again with Henrietta Ball in 1937.

13. CBC Digital Archives, "Egos and Ownership."

14. One can almost imagine that William Butler Yeats, recipient of the Nobel Prize in Literature that same year, was thinking about this "insulin armistice" when he wrote:

> And God would bid his warfare cease,
> Saying all things was well;
> And softly make a rosy peace,
> A peace of heaven with hell
> —W. B. Yeats, "The Rose Of Peace" in *The Rose,* 1893.

15. Sir Frederick Banting died prematurely in February 1941. Dictionary by LaborLawTalk, "Sir Frederick Banting," http://dictionary.laborlawtalk.com/ Frederick_Grant_Banting; Canadian Institutes of Health Research, "Banting's Ties to CIHR," http://www.cihr-irsc.gc.ca/e/25488.html. Banting was making his first-ever transatlantic flight to England. The nature of this trip has never been fully explained; there's been speculation that it had to do with the development of biological/germ warfare weapons to be used in the event that Germany invaded England. Shortly after takeoff from Newfoundland, the Lockheed Hudson bomber carrying Banting crashed. Only the pilot survived; Banting recovered consciousness briefly and, despite serious injuries, was able to attend to the pilot's wounds before dying the next day from hypothermia and exposure. "Aviation in Newfoundland," http://www.angelfire .com/nf/nutting.

16. During the Second World War, Banting became involved in concerns of national defense such as atomic energy, jet aircraft, and even (to the chagrin of some today) germ warfare. He also developed a keen interest in problems related to combat flying, and was one of the first scientists to identify the dangers associated with excessive g-forces. Canada Wide Virtual Science Fair, "Zuckerman L. Blackout," Regina, Sask.: Virtual Science Fair, http://www .virtualsciencefair.org/2006/zuck612/Research.html. These forces caused fighter pilots to occasionally crash after pulling out from steep turns. Banting deduced the reason for this: high-speed acceleration forced blood away from the brain, rendering the pilot unconscious. Aided by Wilbur Franks and others at the University of Toronto, Banting helped construct the "Franks Flying Suit," considered by many to be the first "Anti-G-suit" in aviation history. Ibid. Coincidentally, the Mayo Clinic would play the next major role in the flight suit story.

17. Hormone Foundation, "Estrogen Timeline (1889–1920)" (Chevy Chase, Md.: Hormone Foundation, 2008), http://www.hormone.org/Menopause/ estrogen_timeline/index.cfm.

18. C. A. Rothenberg, "The Rise and Fall of Estrogen Therapy: The History of HRT," http://leda.law.harvard.edu/leda/data/711/Rothenberg05.rtf.

19. Neurotopia, "Menopause," posted February 28, 2006 (no URL provided). It's hard to imagine that pulverized cow ovaries didn't already taste great without the extra flavoring.

20. Hormone Foundation, "Estrogen Timeline."

21. Kendall could not have failed to notice the publication of this work. Would he have been intrigued by it—perhaps intrigued enough to pursue his own investigation into this area? It's doubtful. The Mayo Clinic, with its conservative underpinnings and strong Catholic hospital affiliation, was not likely to become a hotbed for contraception research.

22. E. A. Doisy, C. D. Veler, and S. J. Thayer, "Preparation of Crystalline Ovarian Hormone from Urine of Pregnant Women," *Journal of Biological Chemistry* 86 (1930): 499–509; A. Butenandt and E. von Ziegner, "On the Physiological Effectiveness of Crystallized Female Sex Hormones in the Allen-Doisy Test: Studies on the Female Sex Hormone," *Journal of Physical Chemistry* 188 (1930): 1–10.

23. Google Answers, "*Estrogen*," Google, 2009, http://answers.google.com/ answers/threadview/id/78544.html.

24. Ibid.; W. Frobenius, "Ludwig Fraenkel, Corpus Luteum and Discovery of Progesterone," *Zentralblatt für Gynäkologie* 120, no. 7 (1998): 317–23.

25. Rothenberg, "Rise and Fall of Estrogen Therapy."

26. Ibid.

Vancouver Public Library
Checkout Receipt

Coping with prednisone : (and other cor
Item: 31383082516027
Call No.: 615.36 /84ci
Due Date (MM/DD/YY): 12/12/15

Cortisone
Item: 31383070133478
Call No.: 612.4 K33c
Due Date (MM/DD/YY): 12/12/15

The quest for cortisone /
Item: 31383101420628
Call No.: 615.36 K77q
Due Date (MM/DD/YY): 12/12/15

Total: 3

Item(s) listed due by closing
on date shown.

27. Neurotopia, "Menopause."

28. Like the Canadians, the Germans initially obtained estrogens from human urine—a logical strategy for the country that celebrates Oktoberfest. Rothenberg, "Rise and Fall of Estrogen Therapy."

29. An even more impressive reduction in estrogen cost was made by a team of English chemists led by Charles Dobbs. Concerned that the Germans would dominate the estrogen industry, Dobbs developed diethylstilbestrol (DES) in 1939. Rothenberg, "Rise and Fall of Estrogen Therapy." Incredibly powerful and incredibly cheap, DES was a nonsteroid estrogen that was easy for chemists to make. Dobbs, thinking of the public good, relinquished his patent rights, as was customary at that time. Subsequent sales were brisk. DES eliminated the need to rely on pregnant mares; the estrogen industry was keeping approximately 75,000 mares in small stalls designed for urine collection, and their slaughtered foals had become a multimillion-dollar-a-year delicacy in countries like Belgium, France, and Japan. Animal rights activists welcomed DES as an alternative to the pregnant mare industry. Rothenberg, "Rise and Fall of Estrogen Therapy."

    By 1946 DES was being touted for the prevention of pregnancy complications like hypertension, low birth weight, loss of pregnancy, and others. Two million women were eventually exposed to it. But not everything works the way it's supposed to. Over the next five years it was determined that minimal, if any, clinical benefits were associated with DES, and that potential complications, including some affecting female offspring, were realities.

30. Hormone Foundation, "Estrogen Timeline."

31. In 1934 several groups found themselves competing aggressively to isolate progesterone. Frobenius, "Ludwig Fraenkel," 317–23. That year, Schering Laboratories prepared 20 milligrams of progesterone from almost 1,500 pounds of ovaries (obtained, in turn, from 50,000 sows). Virginia Commonwealth University, "Progesterone and Steroid Hormones" (Richmond: Virginia Commonwealth University, 1996–2008), http://www.people.vcu.edu/~asneden/Progesterone%20and%20steroid%20Hormones.pdf. The extraction and purification of progesterone was an impressive accomplishment, but variability in the starting materials and extraction process caused different batches of progesterone to have wildly different potencies. Before progesterone could become clinically useful, a way to measure its activity was needed. This was provided by a German gynecologist, Dr. Carl Wilhelm Clauberg, who published a description of his progestational assay in 1930. C. Clauberg, "Physiologic und Pathologie der Sexualhormone, im Besonderen des Hormons des Corpus luteum. I. Der biologische Test für das Luteumhormon (das spezielle Hormon des Corpus luteum) am infantilen Kaninchen," *Zentralblatt für Gynäkologie*

54 (1930): 2757–70. Dr. Clauberg's test utilized the proliferating mucous membrane of rabbits and measured the changes produced by administration of progesterone. Known as "Clauberg's Test," it became the gold standard for assessing the potency of progesterone-containing compounds. Unfortunately, the clever Dr. Clauberg turned out to have dangerous political views.

32. Australian Broadcasting Corporation, "Mitchell N. Testosterone: The Many Gandered Hormone," Australian Broadcasting Corporation, 1998, http://www.abc.net.au/science/slab/testost/story.htm#bigt.

33. Ibid.

34. Ibid.

CHAPTER 4. INTRODUCING DR. HENCH

1. H. Anderson, oral communication, June 2007; H. Masson Copeland, *Pill Hill: Growing Up with the Mayo Clinic* (Charlotte, N.C.: Heritage Letterpress, 2004).

2. R. W. Nickerson, "Looking Back: Philip Showalter Hench, Nobel Laureate," *University of Pittsburgh Medical Center Alumni News* (Spring 1993): 14–16.

3. J. Hench, written communication, April 2007.

4. "The Doctor Who Ushered in Cortisone: Philip S. Hench, Rheumatologist," *Medical News and International Report* 3, no. 4 (February 19, 1979).

5. Ibid.

6. One junior colleague, Dr. Howard Polley, described Hench as "dominating in character, a dynamo who worked during the night and slept in installments during the day. A restless extrovert, he talked incessantly—despite a cleft palate, which made his speech very difficult to understand until one mastered its sounds and rhythms. His co-workers reported that after a few weeks one understood it and completely forgot about it. [Hench] never hesitated to call a colleague at 3 o'clock in the morning." Ibid.

7. Nickerson, "Looking Back."

8. Ibid.

9. "Doctor Who Ushered in Cortisone."

10. H. Butt, oral communication, January 2007.

11. Nickerson, "Looking Back."

12. "World of Science Mourns Dr. Hench, Cortisone Pioneer," *Medical World News*, April 16, 1965, 23.

13. Philip S. Hench Walter Reed Yellow Fever Collection, "Citation for Honorary Degree of Doctor of Science for Philip Showalter Hench, October 26

1940" (Charlottesville: Rector and Visitors of the University of Virginia, 1998–2004), http://etext.lib.virginia.edu/etcbin/fever-browse?id=03632001.

14. M. A. Shampo and R. A. Kyle, "Philip S. Hench: 1950 Nobel Laureate," *Mayo Clinic Proceedings* 76 (2001): 1073.

15. Nickerson, "Looking Back."

16. Ibid. The *Titanic* sank in April 1912, the year Hench entered college. The newspapers had reported that "Nearer, My God to Thee" was the final song played by the band as the ship sank. George Behe's Titanic Tidbits, "The Music of the Titanic's Band," http://ourworld.compuserve.com/homepages/carpathia/page3.htm. Hearing it was surely a still-fresh memory that evoked chilling images of a thousand passengers plunging to a cold, icy death. Turning the solemn hymn into a poor man's Scott Joplin arrangement called for some serious irreverence; it must have taken more than a little nerve for him to have played this tune in public at that time.

17. Nickerson, "Looking Back."

18. Not that this close brush with the academic death penalty cured Hench of classroom indiscretions. A few years later, in medical school, Hench would attend pathology lectures given by the famous Dr. Oscar Klotz, a man who, like Addison, was "notorious for inspiring fear and trembling in his students." Ibid. Apparently it was not enough fear or trembling, because one afternoon Hench fell asleep during the lecture. Klotz ordered him to his office. His classmates, now waiting anxiously in the pathology laboratory, feared that Hench was once again facing suspension. There was no need to worry. He returned in a short time with a smile on his face and another near-miss to his credit.

19. BookRags, "World of Biology on Philip Showalter Hench," http://www.bookrags.com/biography/philip-showalter-hench-wob/.

20. F. Françon, "The Exemplary Life of P. S. Hench," *Medical Biology* 43 (1967): 109–23.

21. J. Glyn, "The Discovery and Early Use of Cortisone," *Journal of the Royal Society of Medicine* 91 (October 1998): 513–17.

22. "Prominence" can be a fleeting entity. Aschoff's reputation would become tarnished when he purportedly became a Nazi sympathizer and embraced the new regime. He is credited—or discredited—with having been a major advocate for the acceptance of Nazism within the German medical scientific community. C. R. Prüll, "Pathology and Politics: Ludwig Aschoff (1866–1942) and the German Way in the Third Reich," *History and Philosophy of the Life Sciences* 19, no. 3 (1997): 331–68.

23. Nickerson, "Looking Back."

24. Ibid.

25. Hench's youngest son, John, offers insight into the often unremarkable day-to-day lifestyle at the Hench household. "[My father] had . . . very refined taste—opera, music, etc. He took me to concerts and operas in Rochester and elsewhere. In some cases, like going to the opera, I was given little choice but to accompany my parents, but in retrospect I am grateful because I learned to love classical music. But he also had many more middle-brow tastes and interests. We watched a lot of television together—Ed Sullivan, the quiz and panel shows, Sid Caesar, and the World Series (or, as he called it, the *World Serious*). He went with me to quite a few Rochester Mustang hockey games. We had season tickets for [University of Minnesota] Gopher football games. . . . We'd either drive up to Memorial Stadium or ride in the chartered bus that took other Clinic staff members and their families to the games. When the Twins came to town, we went to a couple of games at the old Metropolitan Stadium." J. Hench, written communication, April 2007.

26. Ibid.

27. Special Collections and Rare Books, Elmer L. Andersen Library, University of Minnesota, Minneapolis, Regents of the University of Minnesota, Twin Cities, and the University Libraries, 2005, "Johnson T. J. P. S. Hench Collection," http://special.lib.umn.edu/findaid/xml/scrb0005.xm.

28. E. H. Rynearson, letter, July 11, 1957, 1 leaf, Mayo Historical Unit, Mayo Foundation, Rochester, Minn.

29. Nickerson, "Looking Back."

30. Philip S. Hench Walter Reed Yellow Fever Collection, "Philip Showalter Hench (1896–1965)" (Charlottesville: Rector and Visitors of the University of Virginia, 1998–2004), http://etext.virginia.edu/healthsci/reed/hench.html.

31. Philip S. Hench Walter Reed Yellow Fever Collection, "Letter from Philip Showalter Hench to Albert E. Truby, October 6, 1940" (Charlottesville: Rector and Visitors of the University of Virginia, 1998–2004), http://etext.lib.virginia.edu/etcbin/fever-browse?id=03622006.

CHAPTER 5. NICE GUYS, SAINTS, ECCENTRICS, AND GENIUSES

1. J. L. Graner, "Leonard Rowntree and the Birth of the Mayo Clinic Research Tradition," *Mayo Clinic Proceedings* 80, no. 7 (July 2005): 920–22.

2. At Hopkins, Rowntree had been an assistant to the world-famous pharmacologist Dr. J. J. Abel. During their time together, they had not only tackled an

assortment of complex pharmacological problems but had somehow found the time to develop the first artificial kidney. Dr. Rowntree was exactly the kind of scientific maven the Mayo brothers needed to build a world-class Department of Medicine for their growing (and still largely surgical) practice.

3. Although Alvarez's contributions toward the Nobel Prize awarded for cortisone would turn out to be relatively minor, his influence on an entirely different kind of Nobel award would be critical.

4. D. P. Steensma, "Luis Walter Alvarez: Another 'Mayo-Trained' Nobel Laureate," *Mayo Clinic Proceedings* 81, no. 2 (February 2006): 241–44, http://www.mayoclinicproceedings.com/content/81/2/241.full.pdf+html.

5. He explained: "I felt that someone should speak out against certain unwise, purely commercial, or heartless types of medical practice. . . . I felt it was my duty to do this, and that if I did not speak up, at some later date I would be ashamed of the fact that I had remained silent." W. C. Alvarez, *Incurable Physician* (Englewood Cliffs, N.J.: Prentice-Hall, 1963), 101.

6. Ibid.

7. Ibid.

8. H. Butt, oral communication, January 2007.

9. Crazy symptoms? Consider this medically controversial self-account from a book Alvarez wrote years later entitled *Allergy of the Nervous System*: "For years I knew I was highly sensitive to chicken, I suffered from what I called 'dumb Monday,' when I was too dull to do much constructive work like writing. Finally, I discovered that bad Mondays were due to the Alvarez family's habit of having chicken for Sunday dinner. . . . My most remarkable personal experience with brain dulling due to food allergy came many years ago when . . . I ate a whole broiled chicken. Next day I had severe diarrhea and with this I became so dulled I could not read with comfort. And that night I had a hallucination of sight, such as I had never had before and haven't had since." Alpha Online: A Division of Environmental Research, "The Brain Center: Food Allergy and the Brain," Sechelt, B.C.: Environmed Research, http://66.51.173.96/brain/allergy_brain.htm.

10. H. Butt, oral communication, January 2007.

11. Ibid.

12. Luis was related by marriage to Mildred Potter, a contralto opera singer said to have performed with Jenny Lind—the aptly named "Swedish Nightingale" and the most famous opera star of her time. Family records indicate that cousin Mildred was the only American to sing at the Metropolitan Opera without having first established a professional career in Europe. Mabel Alvarez (1891–1985) Estate Collection, "Bassett G. Mabel Alvarez (1891–1985): A Personal Memory," http://mabelalvarez.com/about/bassett.htm.

13. Ibid. He arrived in 1888 just as King Kalakaua was recruiting a physician to care for workers in the new plantation industry. Luis took the position and eventually added an interesting side-job—he became the personal physician for the queen of Hawaii. The doctor acquired a significant amount of real estate in downtown Honolulu, and when land prices began to rise in the 1890s, he became an extremely wealthy man.

14. Ibid.

15. At least that was the official story. Luis's daughter, Mabel, told a quite different version of her father's departure from the island. Ibid.

    According to her, a haole (white lady) came to Dr. Alvarez for leprosy testing, and although she appeared healthy her test for the dreaded disease was positive. At that time Hawaiian law mandated that she be placed under quarantine at the leper colony on Molokai, where she would likely spend the rest of her life. She pleaded with Luis to let her return to her home, with the promise that she would voluntarily commit herself should symptoms arise. Kind-hearted Dr. Alvarez agreed to the deception.

    If syphilis, gonorrhea, herpes, and HIV have taught us anything, it is that romance and social diseases don't mix—or mix too well. Unfortunately for Dr. Alvarez, his patient had a lesbian lover who discovered the leprosy status of her partner and "turned her in" to the government health authorities. Luis Alvarez was caught in a political firestorm over his apparent indiscretion and decided it would be best for him to leave paradise at once.

16. Ibid.

17. Walter Alvarez's baby sister Mabel was born in Hawaii on November 28, 1891. Mabel was an excellent student through high school, but truly distinguished herself in art. She was trained artistically as an impressionist, but was molded by trends in modernism. Beautiful, awesomely talented, and adventuresome in spirit, the dark-haired, dark-eyed maven spent more than six decades as a leading American painter. A sensitive and seductive woman—perhaps the very embodiment of estrogen's finest attributes?—she had a soul that was surely at odds with that of her older, more conservative brother, Walter. Mabel Alvarez (1891–1985) Estate Collection, "South, W. Mabel Alvarez," http://mabelalvarez.com/about/south.htm.

18. Steensma, "Luis Walter Alvarez."

19. Alvarez, *Incurable Physician*, 112.

20. A. C. Doyle, "The Adventure of the Missing Three-Quarter" in *The Return of Sherlock Holmes* (1905), 626.

21. eNotes.com, "Alvarez, Luis (1911–1988)," eNotes.com, 2009, http://www.enotes.com/earth-science/alvarez-luis.

22. Steensma, "Luis Walter Alvarez."

23. Ibid.

24. At least Luis realized there were limits to what he should and shouldn't do; there is "no evidence that he did anything that might harm another person, and he was apparently never arrested." Ibid. Now there's every parent's dream-child.

25. H. Masson Copeland, *Pill Hill: Growing Up with the Mayo Clinic* (Charlotte, N.C.: Heritage Letterpress, 2004), 101.

26. Ibid., 102. These two Kendall boys were undoubtedly exceptional climbers (perhaps their father's trip across the Manhattan bridge catwalk reflected a "Spider-Man" gene that the chemist was passing along to his offspring), and they acquired reputations for recklessness by climbing other unprotected edifices about town—for example, the city's water tower in Saint Marys Park, which was easily tall enough to kill them if they fell from it. Some of these climbs led to near disasters, or at least embarrassing moments, especially when less-skilled companions were with them. In her reflections on life in Rochester, Helen Masson Copeland describes an episode that occurred at the Kendalls' cottage on Lake Zumbro: "My brother Stan and Norm Kendall were exploring the space above the rafters at the Kendall's [*sic*] Cedar Beach house one day when Stan, not knowing the flimsy nature of Masonite, stepped off the rafters onto the Masonite ceiling over the porch below. He crashed through, landing in a shower of dust beside the chair where Dr. Kendall sat reading the *New York Times*. A boy falling almost into his lap distracted for a moment Dr. Kendall's attention to the world news. He looked up. 'Norman!' he called. 'You and Stanley are going to have to fix this ceiling.'" Ibid., 162.

27. Ibid., 80. Once Hugh Kendall—who everyone, including his mother, conceded was "most like his father" in looks and disposition—overcame his fixation with death, he focused on a new project: he began showing movies in the attic of his home for five cents a viewing. Ibid., 83. The "Movie King Theatre" included a phonograph (the show opened with a rousing version of "Anchors Aweigh"), homemade curtain, 16mm projector, and a steady supply of movies (those starring Laurel and Hardy or Charlie Chaplin were favorites) and cartoons, often by Disney, that Hugh's mother rented for him from the Kodascope Film Library. The young entrepreneur was "packing them in" at the rate of forty or fifty kids twice a day every Saturday, and making money like a Rockefeller.

28. As Steven Weinreb, the Russell and Mildred Marker Professor of Chemistry at Pennsylvania State University, put it: "There are more stories told about Russell Marker than perhaps any other chemist. Although many of these stories are apocryphal, they are so fascinating that most of us cannot bear to stop repeating them. This is the oral history of our profession that we pass to our

colleagues and our students. They are the campfire stories that bind our profession together." L. Raber, "Steroid Industry Honored," *ACS News* 77, no. 43 (October 25, 1999): 78–80.

29. Ibid.

30. M. Redig, "Yams of Fortune: The (Uncontrolled) Birth of Oral Contraceptives," *Journal of Young Investigators* 6 (2008), http://www.jyi.org/features/ft.php?id=540

31. American Chemical Society, "The Life of Russell Marker" (Washington, D.C.: American Chemical Society, 2004), http://acswebcontent.acs.org/landmarks/marker/life.html.

32. T. Hilchey, "Tadeus Reichstein, 99, Dies: Won Nobel for Cortisone Work," *New York Times* (online edition), August 6, 1996, http://www.nytimes.com/1996/08/06/world/tadeus-reichstein-99-dies-won-nobel-for-cortisone-work.htm; M. Rothschild, "Tadeus Reichstein: 20 July 1897–1 August 1996," *Biographical Memoirs of Fellows of the Royal Society* 45 (1999): 450–67.

33. NNDB: Tracking the Entire World, "Tadeus Reichstein," Soylent Communications, 2009, http://nndb.com/people/371/000128984.

34. Nobelprize.org, "Tadeus Reichstein: The Nobel Prize in Physiology or Medicine 1950," Nobel Web AB, 2009, http://nobelprize.org/nobel_prizes/medicine/laureates/1950/reichstein-bio.html.

35. The backers of these projects got their money's worth. Reichstein discovered that sulfur-containing compounds were important in aroma production, and that these substances could be chemically manufactured and altered. Reichstein eventually found over fifty aromatic compounds in coffee and chicory. Michigan State University Department of Chemistry, "MSU Gallery of Chemists' Photo-Portraits and Mini-Biographies: Tadeus Reichstein (1897–1996)" (East Lansing: Michigan State University, Department of Chemistry, 2008), http://www2.chemistry.msu.edu/portraits/portraitsHH_Detail.asp?HH_LName=Reichstein. His work on aroma-producing substances ended in 1931, at which point he became a full-time research assistant devoted exclusively to academics.

36. He was also doing well socially. In 1927 he married Henriette Louis Quarles Van Ufford, a member of the Dutch Royal Family. BookRags, "World of Biology on Tadeus Reichstein," http://www.bookrags.com/biography/tadeus-reichstein-wob/; Nobelprize.org, "Tadeus Reichstein." Well positioned and well funded are fine. But well connected is better.

37. B. Witkop, "Percy Lavon Julian: April 11, 1899–April 19, 1975" (Washington, D.C.: National Academies Press, Biographical Memoirs), http://www

.nap.edu/html/biomems/pjulian.html/; D. Wyre and J. R. Williams, "Percy Lavon Julian—Life and Legacy," http://www.temple.edu/chemistry/main/faculty/Williams/Percy%20Lavon%20Julian%201.htm.

38. About.com, "Inventors: History of Cortisone," About.com, a part of the New York Times Company, 2009, http://inventors.about.com/library/inventors/blcortisone.htm.

39. Nobelprize.org, "Albert Szent-Györgyi: The Nobel Prize in Physiology or Medicine 1937," Nobel Web AB, 2009, http://nobelprize.org/nobel_prizes/medicine/laureates/1937/szent-gyorgyi-bio.html.

40. American Chemical Society, "Schultz, J. Thinking What No One Had Thought: Albert Szent-Györgyi and the Discovery of Vitamin C" (Washington, D.C.: American Chemical Society, 2009), http://portal.acs.org/portal/acs/corg/content?_nfpb=true&_pageLabel=PP_ARTICLEMAIN&node_id=924&content_id=WPCP_007614&use_sec=true&sec_url_var=region1&uuid.

41. After completing his medical studies he did additional (and very eclectic) postgraduate work: pharmacology in Pozsony, cellular electrophysiology in Prague, and physical chemistry (the same subject over which Russell Marker had angrily left the University of Maryland and nearly wound up becoming a "urine boiler") in Berlin and Hamburg. In 1920 he joined the staff at the University Pharmacology Institute in Leiden, and then from 1922 to 1926 he worked in the Physiology Institute in Groningen, the Netherlands.

42. E. C. Kendall, *Cortisone: Memoirs of a Hormone Hunter* (New York: Charles Scribner's Sons, 1971), 45–46.

43. Ibid.

44. M. A. Shampo and R. A. Kyle, "Edward C. Kendall: Nobel Laureate," *Mayo Clinic Proceedings* 76 (2001): 1188.

45. Glutathione would eventually undergo further studies that would clarify its physiological role. It is a powerful antioxidant found in virtually every cell of the body. Glutathione interacts with enzymes that regulate the metabolism of nutrients, gene expression, DNA and protein synthesis, cellular growth and proliferation, and even some aspects of the immune response. Recent studies suggesting that fruits and vegetables (sources high in dietary glutathione) may protect against certain types of cancer have stimulated today's continuing interest in glutathione research.

CHAPTER 6. 1929 AND THE DECISION TO HUNT FOR CORTISONE

1. Alexander Fleming wasn't feeling depressed either. In 1929 the quiet English scientist was in the process of publishing his soon-to-be famous observations on penicillin. It was work that would eventually lead him to a share of the Nobel Prize in Physiology or Medicine.

2. Biographical Directory of the United States Congress 1774–Present, "Kellogg, Frank Billings (1856–1937)," http://bioguide.congress.gov/scripts/biodisplay.pl?indexK000065.

3. Despite his humble academic background, Kellogg must have been a fairly decent lawyer, because in 1887 he was invited to join a successful law firm in St. Paul run by his cousin, Cushman Kellogg Davis. Frank Kellogg soon made a fortune by looking out for the legal interests of railroads, iron mines, and other manufacturing concerns located in Minnesota. His involvement in the steel industry allowed him to become acquainted with both Andrew Carnegie and John D. Rockefeller. Nobelprize.org, "Frank B. Kellogg: The Nobel Peace Prize 1929," Nobel Web AB, 2009, http://nobelprize.org/nobel_prizes/peace/laureates/1929/Kellogg-bio.html.

4. Ibid.

5. Ibid.

6. U.S. Department of State, "Biography: Frank Billings Kellogg," http://www.state.gov/secretary/former/40829.htm.

7. E. C. Kendall, *Cortisone: Memoirs of a Hormone Hunter* (New York: Charles Scribner's Sons, 1971), 49.

8. Ibid., 48. Although he found Minnesota "not especially interesting," he thought the people were "very nice." R. A. Kyle and M. A. Shampo, "Albert Szent-Györgyi: Nobel Laureate," *Mayo Clinic Proceedings* 75 (2000): 722.

9. F. A. Hartman, K. A. Brownell, and W. E. Hartman, "A Further Study of the Hormone of the Adrenal Cortex," *American Journal of Physiology* 95 (1930): 670–80.

10. W. W. Swingle and J. J. Pfiffner, "Studies on the Adrenal Cortex: I. The Effect of a Lipid Fraction Upon the Life-Span of Adrenalectomized Cats," *American Journal of Physiology* 96, no. 1 (1931), 153–63.

11. "Studies in Addison's Disease: The Muirhead Treatment," *Canadian Medical Association Journal* 15, no. 4 (April 1925): 410–11. It was named after A. L. Muirhead (a professor of pharmacology at the Creighton University School of Medicine in Omaha, Nebraska), who came to the Mayo Clinic suffering from advanced Addison's disease. The professor knew he had a fatal condition; his options were limited. He agreed to try an untested treatment. Muirhead began receiving injections of epinephrine (which is an extract from the adrenal

medulla, not the cortex), took epinephrine suppositories by rectum three times a day, and ate raw adrenal glands with each meal. It was a proctologically and gastronomically punishing form of therapy, but "gratifying improvement resulted." Unfortunately, Muirhead's improvement only lasted for a few months; despite this, the "Muirhead regimen" was tried on other patients.

12. Ibid.

13. L. G. Rowntree, A. R. Kinter, and R. M. Lymburner, "The Use of Swingle and Pfiffner's Cortical Hormone in Addison's Disease: Presentation of Cases," *Staff Meetings of the Mayo Clinic* (August 1930): 216–19.

14. Ibid.

15. This critically ill man, who two days earlier had been unable to keep down water because of retching, now wanted "wieners and sauerkraut" for every meal (no luck there—he had to make do with double portions of the hospital cafeteria beefsteak).

16. Kendall, *Cortisone*, 47.

17. Ibid., 55.

18. Ibid., 46.

19. Ibid.

20. Kendall remembered the conversation a little differently. In his memoirs, he wrote that he and Szent-Györgyi decided that Kendall should pursue studies involving the adrenal cortex after the Hungarian's departure. Rowntree merely provided "stimulation" for that decision. J. L. Graner, "Leonard Rowntree and the Birth of the Mayo Clinic Research Tradition," *Mayo Clinic Proceedings* 80, no. 7 (July 2005): 920–22. In any case, Kendall agreed conceptually with Rowntree. The new focus of the Kendall lab at the Mayo Clinic would be adrenal cortical hormones.

21. Kendall, *Cortisone*, 49.

22. Sir Arthur Conan Doyle, the creator of Sherlock Holmes, died in England on July 7, 1930. Although he made it to the ripe age of seventy-one, he could not escape the final assault of the most ubiquitous human steroid—cholesterol. After building up slowly for many years, the greasy, waxlike material finally plugged Doyle's coronary arteries, giving him crippling angina and eventually killing him with a massive, sudden heart attack. The writer was found dead in his garden, clutching his chest. Sherlockholmesonline.org: The Official Web Site of the Sir Arthur Conan Doyle Literary Estate, "Sir Arthur Conan Doyle biography," Sir Arthur Conan Doyle Literary Estate, 2000, http://www .sherlockholmesonline.org/biography/biography15.htm. His death must have been an emotional passing for Philip Hench, the Holmes devotee who had just returned from his year of study in Europe and was now getting back into the routine of clinical rheumatology practice.

23. J. J. Li, *Laughing Gas, Viagara, and Lipitor: The Human Stories Behind the Drugs We Use* (Oxford: Oxford University Press, 2006), 214.

24. J. H. Glyn, "The Discovery of Cortisone: A Personal Memory," *BMJ* 317, no. 7161 (September 19, 1998): 822A.

CHAPTER 7. ANOTHER KENDALL FALSE START, ANOTHER GREAT ANNOUNCEMENT

1. eSSORTMENT.com, "Dr. Percy L. Julian," eSSORTMENT.com, 2005, http://www.essortment.com/all.percyjuliandr_rmjf.htm.

2. PBS.org, "Nova: Forgotten Genius," WGBH Education Online, 1996–2009, http://www.pbs.org/wgbh/nova/julian/time-nf.html.

3. D. Wyre and J. R. Williams, "Percy Lavon Julian—Life and Legacy," http://www.temple.edu/chemistry/main/faculty/Williams/Percy%20Lavon%20Julian%201.htm.

4. Ibid.

5. Ibid.

6. Ibid.

7. It was no small accomplishment: at the time, Appleton had a law stating that "no Negro should be bedded or boarded in Appleton overnight." Ibid. Julian accepted the job anyway, and the law—an embarrassing leftover from an era of gross intolerance—was soon changed.

8. PBS.org, "A Science Odyssey: People and Discoveries: Percy Julian (1899–1975)," WGBH, 1998, http://www.pbs.org/wgbh/aso/databank/entries/bmjuli .html.

9. Writing in the *Financial Times*, Ian Thomson notes: "The best Sherlock Holmes stories were written before Conan Doyle converted to spiritualism in 1917 and the pseudo-religion eclipsed all other concerns in his later years. A hybrid of mysticism and low church gloom, it flourished amid the bereavement of the first world war in which Conan Doyle lost his adored son Kingsley." Doyle seemed to abandon the logic that made Holmes so popular, promoting schlock over science. "He called for a new science of the paranormal (to be called plasmology) and believed the deceased came to life through luminous voice trumpets." The impact of Doyle's bizarre theological transformation was a major one, affecting his writing "as well as his various love affairs and spiritualist crises." Love affairs? Extramarital activity was clearly out of Dr. Hench's Episcopal comfort zone. Doyle's conservative mother was surely flummoxed by her son's behavior. "As a convert to Anglicanism, Mary Doyle disapproved

of her son's spiritualism—where was the morality in spook-dabbling? His attempts to authenticate photographs of wood nymphs, documented in his non-fiction, *The Coming of the Fairies* (1922), especially pained her as it provoked ridicule. G. K. Chesterton feared Conan Doyle had taken leave of his senses." FT.com: Financial Times, "Arthur Conan Doyle: Review by Ian Thompson," Financial Times, 2009, updated October 6, 2007, http://www .ft.com/cms/5/2/41503bee-70b9–11dc-98fc-0000779fd2ac.html.

10. G. Weaver, *Conan Doyle and the Parson's Son: The George Edalji Case* (New York: Vanguard Press, 2006).

11. The Chronicles of Sir Arthur Conan Doyle, "The George Edalji Case," Perry Internet Consulting, 2009, http://www.siracd.com/life_case1.shtml.

12. E. C. Kendall, *Cortisone: Memoirs of a Hormone Hunter* (New York: Charles Scribner's Sons, 1971), 58.

13. Ibid., 53.

14. D. J. Ingle, *Edward C. Kendall, 1886–1972: A Biographical Memoir* (Washington, D.C.: National Academy of Sciences of the United States, 1975), 258.

15. E. C. Kendall, "Chemical Studies on the Suprarenal Gland: II," *Proceedings of the Staff Meetings of the Mayo Clinic* 7, no. 9 (March 2, 1932): 135–36.

16. Interestingly, a reading of Kendall's memoirs makes the search for lactyl epinephrine sound like a worthy and noble scientific detour. His assistant, Dwight Ingle, looked back on the episode much differently. Ingle was kind in his remarks about Kendall, but clearly thought that the lactyl epinephrine episode was another of his boss's bad ideas gone horribly wrong.

17. Kendall, *Cortisone*, 102.

18. Down the hall from the Kendall operation was a research laboratory run by Ancil Keyes, a recently transplanted Englishman and—metaphorically—one of the brightest bulbs to ever illuminate the Mayo Clinic. Keyes was studying the factors that affected osmotic pressure in various tissues of the body. The experimental work needed to be done in a refrigerated room. Performing these experiments was Keyes's new assistant, Hugh Butt from Virginia, who had just arrived in Rochester that summer and did not yet own a winter coat. The skinny postdoctoral southerner nearly froze in the gigantic walk-in cooler every time he worked inside it. One summer afternoon, while having tea with Dr. Keyes, the kindly Dr. Kendall noticed the young researcher shivering; after figuring out the problem, he brought a warm coat in from home and gave it to the frigid junior scientist from Dixie. H. Butt, oral communication, January 2007.

19. E. C. Kendall, H. L. Mason, B. F. McKenzie, C. S. Myers, and G. A. Koelsche, "Isolation in Crystalline Form of the Hormone Essential to Life from

the Suprarenal Cortex: Its Chemical Nature and Physiologic Properties," *Proceedings of the Staff Meetings of the Mayo Clinic* 9, no. 17 (April 25, 1934): 245–49.

20. Ibid.

21. Ibid.

22. redOrbit.com, "Sellnow G. Flashback: The Kennedy Clan's Rochester Connection," redOrbit.com, 2002–2009, http://www.redorbit.com/news/science/499035/flashback_the_kennedy_clans_rochester_connection/index.html.

23. Kennedy's mental state during this time is reflected in the numerous letters he wrote to family and friends. In public, or in front of his doctors, Jack was a stoic and a wit; in his private letters, he reveals the anxiety and frustration he was experiencing as a sick teenager. "I am suffering terribly out here," he wrote to his best friend. "I now have a gut ache all the time. . . . God what a beating I'm taking. I've lost eight pounds. And still going down. . . . I'm showing them a thing or two. Nobody [is] able to figure what's wrong with me. All they do is talk about what an interesting case." A week later another letter to the same friend shows a definite decline in his fortitude. "I've got something wrong with my intestines. In other words I [expletive] bloody. . . . Yesterday I went through the most harassing experience of my life. . . . [A doctor] stuck an iron tube 12 inches long and 1 inch in diameter up my [backside]. . . . My poor bedraggled rectum is looking at me very reproachfully these days."

Like the good politician he was destined to become, he closed his discourse with a punch. "The reason I'm here is that they may have to cut out my stomach!!!!—the latest news." Questia: The Online Library of Books and Journals, "R. Dallek, Medical Ordeals of JFK: Recent Assessment of Kennedy's Presidency Have Tended to Raise 'Questions of Character'—to View His Administration in the Context of His Sometimes Wayward Personal Behavior," Questia Media America, 2009, http://www.questia.com/PM.qst;jessionid=G8BJYvJgKpYwd2WSwPTCgWvYdvllnHDSXQkyLFWWHnjB4Yzj22v5!639260901?a=o&d=5002448218.

CHAPTER 8. KENDALL STRIKES OUT AGAIN

1. As 1931 rolled around Hemingway was still more or less enamored of F. Scott Fitzgerald ("His talent," waxed Hemingway, "was as natural as the pattern that was made by the dust on a butterfly's wings"; "Simpson's Contemporary Quotations," compiled by James B. Simpson [1988]: number 7403, http://

www.bartleby.com/63/3/7403.html)—but not so infatuated that he couldn't offer a little pointed advice to his literary colleague from Minnesota. "We are all bitched from the start and you especially have to be hurt like hell before you can write seriously," he wrote to Fitzgerald. "But when you get a damned hurt use it—don't cheat with it. Be as faithful to it as a scientist." "Simpson's Contemporary Quotations," compiled by James B. Simpson (1988): number 6902, http://www.bartleby.com/63/2/6209.html.

2. N. Milford, *Zelda: A Biography* (New York: Harper Perennial, 1992), 122.

3. Although Fitzgerald was a Minnesota native who lived in close proximity to Rochester, the Fitzgeralds never played a role in the cortin story. Ernest Hemingway, on the other hand, was destined to become directly involved with two key elements of the tale—the Nobel Prize and the Mayo Clinic.

4. P. S. Hench, "Analgesia Accompanying Hepatitis and Jaundice in Cases of Chronic Arthritis, Fibrositis, and Sciatic Pain," *Proceedings of the Staff Meetings of the Mayo Clinic* 8, no. 28 (July 12, 1933): 430–36; P. S. Hench, "Report of the Fourth Rheumatism Conference," *Proceedings of the Staff Meetings of the Mayo Clinic* 10, no. 39 (September 25, 1935): 615–19.

5. E. C. Kendall, *Cortisone: Memoirs of a Hormone Hunter* (New York: Charles Scribner's Sons, 1971), 65.

6. D. J. Ingle, *Edward C. Kendall, 1886?1972: A Biographical Memoir* (Washington, D.C.: National Academy of Sciences of the United States, 1975), 258.

7. Kendall, *Cortisone*, 66.

8. A. Conan Doyle, *The Adventure of the Retired Colourman*, http://mignon .ddo.jp/assembly/mignon/holmes/colourman.html.

9. E. Regis, "A Cut Lemon Doesn't Turn Brown," *New York Times*, March 6, 1988, http://www.nytimes.com/1988/03/06/books/a-cut-lemon-doesn-t-turn -brown.html?pagewanted=2.

10. J. L. Svirbely and A. Szent-Györgyi, "The Chemical Nature of Vitamin C," *Biochemical Journal* 26 (1932): 865–70.

11. Ibid.

12. This well-deserved scientific recognition did not end the Hungarian's problems. He now faced another huge obstacle. Having used up his supply of hexuronic acid (most of which had been made in Rochester), he had neither reserves of the material nor access to large amounts of adrenal glands from which to extract more. Although he knew it was present in various fruits and vegetables (particularly the type that don't turn brown quickly after you cut them, like lemons and oranges), he'd had little luck extracting sufficient quantities of hexuronic acid from those to which he had access. The solution to this dilemma came in a uniquely Hungarian way. Szent-Györgyi decided to assess the vitamin C content of paprika peppers (Szeged was the paprika

capital of Hungary); it turned out that paprika was an exceedingly rich source of vitamin C. Extracting with a vengeance, his team was able to produce over three pounds of pure crystalline vitamin C—in one week—near the end of 1932. Profiles in Science: National Library of Medicine, "The Albert Szent-Györgyi Papers: Szeged, 1931–1947: Vitamin C, Muscles, and WWII" (Bethesda, Md.: U.S. National Library of Medicine), http://profiles.nlm.nih .gov/WG/Views/Exhibit/narrative/szeged.html.

13. Ibid.

14. It seems Kendall's initially optimistic assessment of his dog experiments was a "false positive" that occurred because the dogs were being fed a high sodium diet, which will temporarily lessen the effects of Addison's disease. Ingle, *Edward C. Kendall*, 259. Kendall, a chemist and not a physiologist, missed the boat on this conclusion. It is yet another instance in which Kendall— ignoring the warning of Sherlock Holmes—would seemingly "twist facts" to suit his theories.

15. Ibid., 256.

16. Ibid., 260.

17. Kendall, *Cortisone*, 65.

18. Ingle, *Edward C. Kendall*, 263.

CHAPTER 9. KENDALL PRESSES ON

1. Mabel Alvarez (1891–1985) Estate Collection, "Bassett G. Mabel Alvarez (1891–1985): A Personal Memory," http://www.mabelalvarez.com/about/ bassett.htm.

2. His patients included Agnes de Mille (Kennicott had not only removed her appendix, but had once been engaged to marry her), Jean Harlow, Edward G. Robinson, Douglas Fairbanks, Mary Pickford, and a host of others.

3. One of her friends at the time, model Adrienne Tytla, summed up the unspoken feelings that Mabel's friends had long harbored regarding the uncomfortable one-way relationship: "I knew Bob was gay the first time I met him," she explained; "I can't imagine Mabel was so naïve that she took that long to figure it out." Mabel Alvarez (1891–1985) Estate Collection, "Bassett G. Mabel Alvarez (1891–1985): A Personal Memory," http://www .mabelalvarez.com/about/bassett.htm.

4. Mabel Alvarez (1891–1985) Estate Collection, "South, W. Mabel Alvarez."

5. D. J. Ingle, *Edward C. Kendall, 1886–1972: A Biographical Memoir* (Washing-ton, D.C.: National Academy of Sciences of the United States, 1975), 261.

6. E. C. Kendall, *Cortisone: Memoirs of a Hormone Hunter* (New York: Charles Scribner's Sons, 1971), 64.

7. Ibid., 69.

8. Ibid.

9. Michigan State University Department of Chemistry, "MSU Gallery of Chemists' Photo-Portraits and Mini-Biographies: Tadeus Reichstein (1897–1996)" (East Lansing: Michigan State University, Department of Chemistry, 2008), http://www2.chemistry.msu.edu/portraits/portraitsHH _Detail.asp?HH_LName=Reichstein.

10. American Chemical Society, "The Life of Russell Marker" (Washington, D.C.: American Chemical Society, 2004), http://acswebcontent.acs.org/ landmarks/marker/life.html.

11. Kendall, *Cortisone*, 69–71.

12. Ibid., 71.

CHAPTER 10. SCORE: SZENT-GYÖRGYI–1; KENDALL–0

1. Profiles in Science: National Library of Medicine, "The Albert Szent-Györgyi Papers: Szeged, 1931–1947: Vitamin C, Muscles, and WWII" (Bethesda, Md.: U.S. National Library of Medicine), http://profiles.nlm.nih.gov/WG/ Views/Exhibit/narrative/szeged.html.

2. Exactly what role Alvarez played in Kennedy's care remains cloaked in medical privacy, but it is likely that they had a strong doctor-patient relationship. There was certainly a social relationship; Kennedy would, from time to time, be a guest at Alvarez's home, particularly during the years to come. D. P. Steensma, "Luis Walter Alvarez: Another 'Mayo-Trained' Nobel Laureate," *Mayo Clinic Proceedings* 81, no. 2 (February 2006): 241–44.

3. Questia: The Online Library of Books and Journals, "R. Dallek, Medical Ordeals of JFK: Recent Assessment of Kennedy's Presidency Have Tended to Raise 'Questions of Character'—to View His Administration in the Context of His Sometimes Wayward Personal Behavior," Questia Media America, 2009, http::://www.questia.com/PM.qst;jsessionid=G8BJYvJgKpY wd2WSwPTCgWvYdvllnHDSXQkyLFWWHnjB4Yzj22v5!639260901?a =o&d=5002448218.

4. The possibility that he might actually die, while remote, seemed—at least to JFK himself—very real. Kennedy tried to keep a cavalier attitude: "Took a peek at my chart yesterday and could see that they were mentally measuring me for a coffin. Eat drink and make Olive [his girlfriend], as tomorrow or

next week we attend my funeral. I think the Rockefeller Institute may take my case . . . flash—they are going to stick that tube up my ass again as they did at Mayo." Ibid.

5. Online NewsHour, "Online Focus: President Kennedy's Health Secrets," MacNeil/Lehrer Productions, 1996–2009, http://www.pbs.org/newshour/bb/health/july-dec02/jfk_11–18.html.

6. Questia: The Online Library of Books and Journals, "R. Dallek, Medical Ordeals of JFK."

7. J. W. Ferrebee, D. Parker, W. H. Carnes, M. K. Gerity, D. W. Atchley, and R. F. Loeb, "Certain Effects of Desoxycorticosterone: The Development of 'Diabetes Insipidus' and the Replacement of Muscle Potassium by Sodium in Normal Dogs," *American Journal of Physiology* 135, no. 1 (November 30, 1941): 230–37.

8. C. Healy and G. Guider, "DOCA-Salt Induced Myocardial Sensitization to Ventricular Fibrillation by Isoprenaline in Rats: Role of the Autonomic Nervous System," *Journal of Autonomic Pharmacology* 5, no. 4 (December 1985): 271–78.

9. Questia: The Online Library of Books and Journals, "R. Dallek, Medical Ordeals of JFK."

10. Ibid.

11. Ibid.

12. George Thorn of the Harvard Medical School was, at that time, the world leader in the clinical use of DOCA. With Harvard located practically in Kennedy's backyard, why go to Mayo to obtain DOCA? Perhaps he went there because Kendall and his group at Mayo were actively involved in the clinical use of DOCA for the treatment of various types of disease. Dr. Charles Code of Mayo had designed the type of DOCA tablets Kennedy used—bean-size pellets containing a mixture of desoxycorticosterone and beeswax that could be inserted under the skin. The water-resistant implants provided a slow release of the substance and ensured its sustained activity over days or weeks. E. C. Kendall, *Cortisone: Memoirs of a Hormone Hunter* (New York: Charles Scribner's Sons, 1971), 81.

13. Hench had picked a good time to expand his interest in Sherlock Holmes; 1938 was shaping up to be an excellent year for Holmes fans. The *Adventures of Robin Hood* with Errol Flynn and Basil Rathbone was a smash hit and further confirmed Rathbone's star status. The veteran actor was now working on the first (*The Adventures of Sherlock Holmes*) of what would prove to be many successful Sherlock Holmes films. The soon-to-be-released big-screen adaptation of the Sherlock Holmes story would vault the fictional detective solidly into the general public's eye.

14. Philip S. Hench Walter Reed Yellow Fever Collection, "Letter from Philip Showalter Hench to John J. Moran, April 21, 1939" (Charlottesville: Rector and Visitors of the University of Virginia, 1998–2004), http://yellowfever.lib .virginia.edu/reed/story.html.

15. Ibid.

16. M. Redig, "Yams of Fortune: The (Uncontrolled) Birth of Oral Contraceptives," *Journal of Young Investigators* 6 (2008), http://www.jyi.org./volumes/ volume6/issue7/features/redig.html.

CHAPTER 11. TRANSITIONS AND TRAVELS

1. Profiles in Science: National Library of Medicine, "The Albert Szent-Györgyi Papers: Szeged, 1931–1947: Vitamin C, Muscles, and WWII" (Bethesda, Md.: U.S. National Library of Medicine), http://profiles.nlm.nih.gov/WG/ Views/Exhibit/narrative/szeged.html.

2. Nobelprize.org, "Albert Szent-Györgyi: The Nobel Prize in Physiology or Medicine 1937," Nobel Web AB, 2009, http://nobelprize.org/nobel_prizes/ medicine/laureates/1937/szent-gyorgyi-bio.html.

3. R. A. Kyle and M. A. Shampo, "Albert Szent-Györgyi: Nobel Laureate," *Mayo Clinic Proceedings* 75 (2000): 722.

4. U.S. National Library of Medicine, "Albert Szent-Györgyi Papers Now Available on Profiles in Science" (Bethesda, Md.: U.S. National Library of Medicine), updated June 15, 2005, http://www.nlm.nih.gov/news/press _releases/szent_gyorgyi_PR05.html.

5. Szent-Györgyi's sudden and surprising popularity offered an ironic chance at redemption for a man who was purported to have gotten out of the army in the previous world war by shooting himself. Wikipedia, the Free Encyclopedia, "Albert Szent-Györgyi," Wikimedia Foundation, updated June 7, 2009, http://en.wikipedia.org/wiki/Albert_Szent-Gy%C3%B6rgyi.

6. The Quotations Page, "Quotations by Author: Albert Szent-Györgyi (1893–1986)," QuotationsPage.com and Michael Moncur, 1994–2007, http:// www.quotationspage.com/quotes/Albert_Szent-Gyorgyi/.

7. R. Dallek, "The Medical Ordeals of JFK," *Atlantic Monthly*, December 2002, 49–61, http://www.theatlantic.com/doc/200212/dallek-jfk/3.

8. Later that year Kennedy slipped under the therapeutic spell of Dr. William Murphy, a Harvard physician who had won the Nobel Prize in 1934 after discovering a treatment for pernicious anemia (the same condition Addison was originally investigating when he discovered "Addison's disease"). Ibid.

Murphy's miracle treatment for pernicious anemia involved the use of liver extracts (which contained vitamin B12 and folic acid); the Nobel laureate had an unhealthy faith in the restorative power of liver extracts for conditions like pernicious anemia, and he was optimistic that Kennedy would respond favorably to the proprietary pâté panacea he'd prepared. Kennedy received his liver injections with the highest of hopes and expectations. Unfortunately, but in retrospect not unpredictably, the liver extracts didn't help him. He remained wracked by episodes of colon spasm, abdominal pain, and weight loss.

9. Ibid.

10. Mayo Clinic, "Dr. Charles H. Mayo Dies," Mayo Foundation for Medical Education and Research, 2001–2007, http://www.mayoclinic.org/tradition-heritage-artifacts/images/65–2-lg.jpg. When it came to lung diseases, most people weren't so discriminating about types; pneumonia was simply "the old man's friend."

11. Ibid.

12. More soldiers had died from yellow fever during the Spanish-American War than died from combat. Epidemics of the fever occasionally hit the U.S. mainland, reaching as far north as Philadelphia (and causing devastating death). By 1900 there were estimates of more than 300,000 cases of yellow fever in the United States, with a mortality rate of 40–85 percent. Until "yellow jack" was tamed, work in the tropics (including the construction of the badly needed Panama Canal) had been impractical.

13. *Yellow Jack*, released in 1938, was directed by George B. Seitz and starred Robert Montgomery, Lewis Stone, and Andy Devine. Henry Hull played Dr. Jesse Lazear, the Washington and Jefferson alumnus at the heart of the upcoming Founder's Day celebration. Hench saw it in the theater and afterward provided this critique: "Unfortunately they put too much comedy in it, and there were only a few places where it was adequately dramatized with sufficient dignity. . . . Nevertheless, it was an interesting movie, even if not too accurate." Philip S. Hench Walter Reed Yellow Fever Collection, "Letter from Philip Showalter Hench to John J. Moran, July 13, 1938" (Charlottesville: Rector and Visitors of the University of Virginia, 1998–2004), http://yellowfever.lib.virginia.edu/reed/story.html).

14. Philip S. Hench Walter Reed Yellow Fever Collection, "Letter from Philip Showalter Hench to Ralph Hutchison, December 10, 1939" (Charlottesville: Rector and Visitors of the University of Virginia, 1998–2004), http://yellowfever.lib.virginia.edu/reed/story.html.

15. This suggestion turned out to be prophetic; Hench wrote to Hutchison: "last night I put one man (Domagk) on my list, and this morning in the paper, I noticed he had received the Nobel Prize. That shows how good a picker

I am." Philip S. Hench Walter Reed Yellow Fever Collection, "Letter from Philip Showalter Hench to Ralph Hutchison, November 3, 1939" (Charlottesville: Rector and Visitors of the University of Virginia, 1998–2004), http://yellowfever.lib.virginia.edu/reed/story.html). Hitler ordered Domagk to refuse his prize.

Perhaps even more prophetic was a conversation taking place in Stockholm at the time. J. Grafton Love, the Mayo Clinic's preeminent neurosurgeon, was visiting Henning Waldenstrom, a friend of his at the Karolinska Institute (the organization that awards the scientific Nobel prizes each year). The Swedish doctor had met Philip Hench on several occasions; he asked Dr. Love, "What's Phil Hench doing" these days? Love answered that it was work involving the reversibility of rheumatoid arthritis as a result of jaundice, pregnancy, and other stressors. After a moment of quiet thought, Waldenstrom opined that the subject sounded "very interesting and might be worth a (Nobel) prize . . . someday." H. F. Polley, unpublished interview, 1975, Mayo Clinic Archives.

16. E. C. Kendall, *Cortisone: Memoirs of a Hormone Hunter* (New York: Charles Scribner's Sons, 1971), 80.

CHAPTER 12. WAR LOOMS

1. The History Place: Great Speeches Collection, "Neville Chamberlain Speech: On the Nazi Invasion of Poland—September 1, 1939," http://www.history place.com/speeches/chamberlain.htm.
2. D. J. Ingle, *Edward C. Kendall, 1886–1972: A Biographical Memoir* (Washington, D.C.: National Academy of Sciences of the United States, 1975), 266.
3. E. C. Kendall, *Cortisone: Memoirs of a Hormone Hunter* (New York: Charles Scribner's Sons, 1971), 99.
4. Ibid., 100.
5. After a year in Hawaii pining over the loss of her beau, Mabel returned to Los Angeles, and in 1941 her work was featured in a one-woman show at the Los Angeles art museum. The exhibit, which featured portraits of native Hawaiians, became extremely popular on the West Coast.
6. Luis Walter Alvarez Biography (1911–1988), NetIndustries, 2008, http://74.125.95.132/search?q=cache:luM82WTks0YJ:www.madehow.com/inventorbios/7/Luis-Walter-Alvarez.html.
7. Luis Alvarez's flight to MIT in Cambridge, Massachusetts, is the subject of a famous Mayo Clinic anecdote. According to legend, Luis's efforts to report to

Cambridge for research duty were hampered when his plane was unable to land in Boston because of bad weather. During the era before terrorist precautions, Luis asked for—and was granted—permission to speak with the pilots as they flew in a holding pattern around Boston. While visiting the cockpit he asked about the type of information needed to navigate in adverse conditions like these. After studying the primitive radar on board, Luis offered his own interpretation of the crude signals and helped the pilots bring the plane home safely; they, in turn, provided Luis with the substrate for another "prize wild idea."

8. Accessmylibrary.com, "Alvarez, Luis (1911–1988)" (Gale, 2002; Thomson Gale, a part of the Thomson Corporation, 2005; World of Earth Science, January 1, 2003), http://www.accessmylibrary.com/coms2/summary _0193–13679_ITM; R. L. Garwin, "Memorial Tribute for Luis W. Alvarez, May 1, 1991," Memorial Tribute for Luis W. Alvarez, in Memorial Tributes, National Academy of Engineering, vol. 5 (Washington, D.C.: National Academy Press, 1992), Garwin Archive, http://www.fas.org/rlg/alvarez.htm.

9. The Hemingway Cookbook, "K. Dupuis, Homing to the Stream: Ernest Hemingway in Cuba," DEE WHY BOOKS, 1999–2002; © Kelley Dupuis, http://www.deewhybooks.com.au/Magazine3/ReadingList.htm.

10. Ibid.

11. In May 1900 forty-nine-year-old Major Walter Reed, a graduate of the University of Virginia Medical School and a "student of physiology" at Johns Hopkins, was appointed by the army to a board whose purpose was to study yellow fever in Cuba. Also serving on this board was Major Jesse W. Lazear (the previously discussed graduate of Washington and Jefferson College), a physician with expertise in infectious disease. Reed oversaw a project to investigate the theory of Dr. Carlos Juan Finlay, a Havana physician who believed that mosquitoes were the cause of yellow fever. Because there was no animal model for yellow fever, studies on the disease had to be performed with humans. In a series of elaborate experiments, certain volunteers allowed themselves to be bitten by mosquitoes. Other volunteers spent days locked up in small buildings containing the clothing, bedding, and so forth of yellow fever victims (but with no mosquitoes) to show that the disease could not be transmitted in the absence of a bug bite. Dr. Lazear allowed himself to be intentionally bitten by an infected mosquito and developed yellow fever with its "black vomit," jaundice, and delirium. He subsequently died. Because of Lazear's heroic contribution—and the contributions made by other volunteers who risked or incurred infection—Walter Reed was able to prove convincingly that yellow fever could be prevented by controlling mosquitoes. The site upon which these experiments were conducted was renamed Camp Lazear in honor of the fallen investigator. Now abandoned

for almost forty years, Philip Hench sought to visit it and consider means by which it might be restored.

12. Philip S. Hench Walter Reed Yellow Fever Collection, "Letter from Philip Showalter Hench to John J. Moran, March 12, 1940" (Charlottesville: Rector and Visitors of the University of Virginia, 1998–2004), http://yellowfever.lib .virginia.edu/reed/story.html.

13. Philip S. Hench Walter Reed Yellow Fever Collection, "Letter from Philip Showalter Hench to Ralph Hutchison, May 6, 1940" (Charlottesville: Rector and Visitors of the University of Virginia, 1998–2004), http://yellowfever.lib .virginia.edu/reed/story.html.

14. Ibid.

CHAPTER 13. HENCH MEETS KENDALL

1. D. P. Steensma, "Luis Walter Alvarez: Another 'Mayo-Trained' Nobel Laureate," *Mayo Clinic Proceedings* 81, no. 2 (February 2006): 241–44.

2. Ibid.

3. Did his family connection to Will Mayo contribute to Dr. Walters's rise through the clinic's hierarchy? Walter Alvarez, writing in his memoirs, raises this possibility when he notes of Dr. Will that "only rarely did he pick an unsuitable man and then usually it was due to bad advice with a touch of nepotism." Was Walter Alvarez tossing this barb at Dr. Walters—perhaps because of the subsequent dismal clinical course his son took after the great surgeon's operation? W. C. Alvarez, *Incurable Physician: An Autobiography* (New Jersey: Prentice-Hall, 1963), 143.

4. Steensma, "Luis Walter Alvarez."

5. H. R. Butt, E. V. Allen, and J. L. Bollman, "A Preparation from Spoiled Sweet Clover [3,3'-methylene-bis(4-hydroxycoumarin)] Which Prolongs Coagulation and ProthrombinTime of the Blood: Preliminary Report of Experimental and Clinical Studies," *Proceedings of the Staff Meetings of the Mayo Clinic* 16 (1941): 388–95.

6. Philip S. Hench Walter Reed Yellow Fever Collection, "Letter from Philip Showalter Hench to His Parents, June 9, 1941" (Charlottesville: Rector and Visitors of the University of Virginia, 1998–2004), http://etext.lib.virginia .edu/etcbin/ot2www-fever?specfile=/web/data/newreed/reed.02w&ac.

7. H. F. Polley and C. H. Slocumb, unpublished notes, Mayo Clinic Archives, Rochester, MN.

8. Ibid.

9. Ibid.

10. Ibid. Polley may have had a legitimate reason to speculate about Hench and Kendall's backroom conversations. Years later, but still before the first experiments involving compound E and rheumatoid arthritis were carried out, Polley would consider trying this experiment himself. "In the spring of 1948 Haddon Carryer [a Mayo graduate student] told Dr. Polley about his animal experiments with compound E in guinea pigs. Carryer was investigating allergy at the time for his PhD thesis and he suggested 'you ought to try compound E for rheumatoid arthritis.'" Dr. Polley was busy studying for the American Board of Internal Medicine exams, and he "never forgot it but never got around to it." If Kendall and Hench had not discussed the use of compound E in rheumatoid arthritis in 1941, as they both claimed, it's possible that Carryer and Polley could have claimed that they came up with the idea first in 1948.

11. Philip S. Hench Walter Reed Yellow Fever Collection, "Letter from Philip Showalter Hench to General Truby, October 16, 1941" (Charlottesville: Rector and Visitors of the University of Virginia, 1998–2004), http://etext.lib.virginia.edu/etcbin/ot2www-fever?specfile=/web/data/newreed/reed.02w&ac.

12. Philip S. Hench Walter Reed Yellow Fever Collection, "Letter from Philip Showalter Hench to Ralph Cooper Hutchison, December 13, 1940" (Charlottesville: Rector and Visitors of the University of Virginia, 1998–2004), http://etext.lib.virginia.edu/etcbin/ot2www-fever?specfile=/web/data/newreed/reed.02w&ac.

13. Ibid.

14. Philip S. Hench Walter Reed Yellow Fever Collection, "Letter from Ralph Cooper Hutchison to Philip Showalter Hench, December 20, 1940" (Charlottesville: Rector and Visitors of the University of Virginia, 1998–2004), http://etext.lib.virginia.edu/etcbin/ot2www-fever?specfile=/web/data/newreed/reed.02w&ac.

15. Ibid.

16. Philip S. Hench Walter Reed Yellow Fever Collection, "Telegram from Philip Showalter Hench to Ralph Cooper Hutchison, December 23, 1940" (Charlottesville: Rector and Visitors of the University of Virginia, 1998–2004), http://etext.lib.virginia.edu/etcbin/ot2www-fever?specfile=/web/data/newreed/reed.02w&ac.

17. Philip S. Hench Walter Reed Yellow Fever Collection, "Telegram from Ralph Cooper Hutchison to Philip Showalter Hench, December 24, 1940" (Charlottesville: Rector and Visitors of the University of Virginia, 1998–2004), http://etext.lib.virginia.edu/etcbin/ot2www-fever?specfile=/web/data/newreed/reed.02w&ac.

18. Ibid.

19. "R. E. Marker, Expert on Hormones, Dies at 92," *New York Times Obituaries*, March 9, 1995, New York Times Company, 2009, http://www/nytimes .com/1995/03/09/obituaries/r-e-marker-expert-on-hormones-dies-at-92.

20. Thus proving Holmes's assessment regarding the worthlessness of "local aid."

CHAPTER 14. WORLD WAR II AND
MILITARY STEROID RESEARCH

1. R. Beamish and R. Ian, "The Spectre of Steroids: Nazi Propaganda, Cold War Anxiety and Patriarchal Paternalism," *International Journal of the History of Sport* 22, no. 5 (September 2005): 777–95; M. Hill, "What's Wrong with Steroids?" *Baltimore Sun*, April 2, 2006, http://www.baltimoresun.com/ sports/olympics/.

2. There is some truth to the oft-made claim that the Nazis "discovered" anabolic steroids—Germany's Adolf Butenandt was the first to isolate androsterone (an anabolic steroid) in the early 1930s; he extracted minuscule amounts from thousands of gallons of urine (obtained from Bavarian policemen). In 1935 Butenandt synthesized a close relative of androsterone—testosterone—from a precursor called trans-dehydro-androsterone. He was awarded the 1939 Nobel Prize in Chemistry for this achievement (along with Leopold Ruzicka, a Swiss professor who performed similar work independently), although he, like Gerhard Domagk, was not able to accept the award until later (1949) because of the war and Hitler's disdain for the Nobel Prize.

3. E. O. Lemons, *The Third Reich, A Revolution of Ideological Inhumanity*, Vol. 2, *Death Mask of Humanity* (Morrisville: Everette O. Lemons, 2006), 381; I. Gutman and M. Berenbaum, eds., *Anatomy of the Auschwitz Death Camp* (Bloomington: Indiana University Press, 1998), 306; Jewish Virtual Library, "Carl Clauberg (1898–1957)," American-Israeli Cooperative Enterprise, 2009, http://www.jewishvirtuallibrary.org/jsource/biography/Clauberg.html.

4. Law Reports of Trials of War Criminals: Selected and Prepared by the United Nations War Crimes Commission, vol. 7 (London: His Majesty's Stationery Office, 1948), 15–16, 33.

5. Ibid.

6. Jewish Virtual Library, "Carl Clauberg."

7. Ibid.

8. Clauberg was never adequately punished for his prison camp inhumanity, not that it would ever be possible to adequately punish crimes like this. Clauberg

escaped capture at Auschwitz by fleeing the advancing Soviet army. He moved to Ravensbruck, where he continued experimenting until the Soviets captured him at that location in 1945. He was tried in the Soviet Union in 1948 and sentenced to twenty-three years in prison. As a result of a prisoner exchange between the Soviet Union and West Germany, Clauberg was released seven years later in 1955. His hubris was undeterred; upon his return to West Germany he boasted of his "scientific achievements" in the camps. Not surprisingly, there were public expressions of protest from survivors and other offended parties. He was rearrested and imprisoned while awaiting retrial. Officially, he died of a "heart attack" in his cell in 1957, but his death is often described as being "mysterious" in cause.

9. The Hunt for Danish KZ, "Carl Peter Vaernet Timeline Based on Available Sources," IHWO, 1998–2002, http://users.cybercity.dk/~dk012530/hunt_for_danish_kz.htm.

10. Ibid.

11. Ibid.

12. As his interest in treating homosexuality increased, Vaernet's desire to continue practicing traditional medicine declined. He sold his clinic in Copenhagen to the Germans, who were now occupying the city. It was later blown up by the Danish resistance movement. Ibid.

13. Ibid.

14. The Memorial Hall: The Gay Holocaust—Nazi Criminals, "6.2.2—Dr. Carl Peter Vaernet (1893–1965)," Matt and Andrej Koymasky, 1997–2004, updated March 7, 2004, http://andrejkoymasky.com/mem/holocaust/06/h006n.html.

15. Hunt for Danish KZ, "Carl Peter Vaernet Timeline."

16. The unfortunately named Brigadier General Helmut Poppendick.

17. W. Röll, "Homosexual Inmates in the Buchenwald Concentration Camp," *Journal of Homosexuality* 31, no. 4 (1996): 1–28.

18. Hunt for Danish KZ, "Carl Peter Vaernet Timeline"; Memorial Hall: The Gay Holocaust—Nazi Criminals, "6.2.2—Dr. Carl Peter Vaernet."

19. Memorial Hall: The Gay Holocaust—Nazi Criminals, "6.2.2—Dr. Carl Peter Vaernet"; D. A. Hackett, *The Buchenwald Report* (Boulder, Colo.: Westview Press, 1997), 79.

20. Hackett, *Buchenwald Report*, 79.

21. Even more so than Clauberg, Vaernet cheated his rightful fate immediately after the war. He was captured and placed in a British prisoner of war camp in Copenhagen in 1945. For reasons that are unclear, the British turned him over to the Danish authorities, who simply lost track of him. Vaernet apparently claimed to suffer from a type of heart trouble that could only be

treated in Sweden at that time; he was, astonishingly, allowed by the Danish government to travel to Stockholm for medical care. Once there, he contacted a Nazi underground network and managed to escape to Argentina. On November 19, 1947, a Copenhagen newspaper reported that Dr. Vaernet was working in the Buenos Aires Health Department. He eventually opened a private clinic, but because he never mastered the Spanish language he couldn't communicate with his patients and was therefore unsuccessful as a physician. But Vaernet had escaped justice, and the only solace one could take was that he lived in constant terror of being discovered and punished for his past crimes. The Hunt for Danish KZ, "Outrage! Letter to Prime Minister Rasmussen of Denmark From Peter Tatchell," IHWO, 1998–2002, http://users .cybercity.dk/~dk012530/hunt_for_danish_kz.htm.

Or perhaps justice was merely late in catching up with Vaernet. In 1955 the doctor was hit by a taxi and severely injured; at least fifteen of his bones were broken. He became physically dependent upon his wife, Gurli, to take care of him. Unfortunately for Vaernet, a few months after his injury she accidentally touched an electric trolley-train power line and was electrocuted. Alone and crippled, Vaernet spent the next one and a half years in the hospital, during which time he exhausted his money and became a pauper.

22. Law Reports of Trials of War Criminals: Selected and Prepared by the United Nations War Crimes Commission, Vol. 7, 15–16, 33.

23. Jewish Virtual Library, "Dachau High-Altitude Experiments Photo (U.S. Holocaust Memorial Museum)," American-Israeli Cooperative Enterprise, 2009, http://www.jewishvirtuallibrary.org/jsource/Holocaust/altexphoto.html.

24. J. Stepanek, "Aviation Medicine at Mayo," *Minnesota Medicine* 86, no. 9 (September 2003): 44–46.

25. Ibid.

26. Charles Lindbergh: An American Aviator, "Mayo Unveils Classified Research Involving Charles Lindbergh during World War II," CharlesLindbergh.com, 1998–2007, http://www.charleslindbergh.com/press/mayo.asp.

27. Using the low-pressure test chamber, scientists at the clinic developed another technique that enabled pilots to tolerate high-altitude flight—breathing 100 percent oxygen before flying to altitudes greater than 25,000 feet (dissolved nitrogen causes high-altitude sickness, much like it causes "the bends" when divers surface too rapidly; oxygen prebreathing eliminates this nitrogen from the blood). Approaches like the "bail-out bottle" and "oxygen prebreathing" remain staples of high-altitude aviation today.

28. N. Berlinger, "The War against Gravity," *Invention and Technology Magazine* 20, no. 4 (Spring 2005), http://www.americanheritage.com/articles/magazine/ it/2005/4/2005_4_18.shtml.

29. Ibid.
30. "Canadian-Invented Flying Suit Prevents 'Blackout' of Pilots," *Maple Leaf*, December 23, 1944; "Hall of Fame: Dr. Wilbur Franks," Banting Research Foundation, 2003, http://www.utoronto.ca/bantresf/HallofFame/Franks.html.
31. J. B. A. MacLeod, "Frederick G. Banting: Giving Prospects for Life from the Past to the New Millennium," *Archives of Surgery* 141, no. 7 (July 2006): 705–7.
32. Berlinger, "War against Gravity."
33. Ibid.
34. Ibid.
35. Wood later acknowledged that pilots really didn't like his suit any better than the one developed by Franks and Banting; many claimed they would rather black out than wear it. Wood dismissed that assessment as machismo. "Pilots are just that way," he said. "They don't think they need anything." Ibid. The data proved otherwise. The new flight suit was implemented in the fall of 1944, and the subsequent results were dramatic. Fighter pilots in Australia discovered that they could make pursuing planes crash just by putting their own plane through high-g maneuvers that caused the enemy to lose consciousness if they attempted the same turns. P51 Mustang flyers in the Eighth Air Force demonstrated that pilots wearing "Anti-G suits" shot down sixty-seven enemy planes per 1,000 flying hours, compared to thirty-three planes for those who did not wear them. Ibid. Ultimately, the Germans and Japanese paid a supreme price for their failure to develop effective "Anti-G suits"—and may have lost the air war because of this omission.

CHAPTER 15. PLANTS, POLITICIANS, AND MORE PESSIMISM

1. "The Life of Percy Lavon Julian '20" (Greencastle, Ind.: DePauw University, 2000–2009), http://www.depauw.edu/news/?id=22969.
2. L. Raber, "Steroid Industry Honored: International Historic Chemical Landmark Acclaims Success of Mexican Steroid Industry and a U.S. Chemist Who Made It Possible," *Journal of the Mexican Chemical Society* 43, no. 6 (1999): 235–37.
3. "R. E. Marker, Expert on Hormones, Dies at 92," *New York Times Obituaries*, March 9, 1995, New York Times Company, 2009, http://www.nytimes.com/1995/03/09/obituaries/r-e-marker-expert-on-hormones-dies-at-92.
4. It's not clear why Marker was unable to garner support for his plans—perhaps it was just another example of his poor interpersonal communication skills

(like the ones that caused him to walk away from the University of Maryland without his PhD). For whatever reason, he failed at Sherlock Holmes's aphorism to "make people believe" in what he had done.

5. "Mexican Yams—Cortisone, Steroids, and Oral Contraceptives," www.botgard .ucla.edu/html/botanytextbooks/economicbotany/Dioscoreamed/.

6. Raber, "Steroid Industry Honored," 235–37.

7. M. Redig, "Yams of Fortune: The (Uncontrolled) Birth of Oral Contraceptives," *Journal of Young Investigators* 6, no. 7 (February 2003), http://www.jyi .org./volumes/volume6/issue7/features/redig.html.

8. Somolo and Lehmann would later regret their failure to learn more about Marker's process for making progesterone.

9. "R. E. Marker, Expert on Hormones, Dies at 92."

10. Redig, "Yams of Fortune."

11. R. Dallek, "The Medical Ordeals of JFK," *Atlantic Monthly*, December 2002, 49–61, http://www.theatlantic.com/doc/200212/dallek-jfk/4.

12. John Kennedy: Hospitalization Chronology, "No. 17—late 1940," DoctorZebra .com, 2000–2009, http://www.doctorzebra.com/prez/z_x35hospitalizations_g .htm.

13. redOrbit.com, "Sellnow G. Flashback: The Kennedy Clan's Rochester Connection," redOrbit.com, 2002–2009, http://www.redorbit.com/news/ science/499035/flashback_the_kennedy_clans_rochester_connection/index .html.

14. Book review: *Why England Slept*, by John F. Kennedy, Tao Yue, 1997–2007, updated February 17, 2007, http://www.taoyue.com/books/why-england -slept.html.

15. Dallek, "Medical Ordeals of JFK."

16. His father, Joseph Kennedy, noted on JFK's return from the war that "we found him in reasonable good shape when he returned . . . but the doctors at Mayo don't entirely agree with me on this diagnosis." Over the next couple of years JFK was also seen at other medical facilities for his various ailments, but no cures were forthcoming. redOrbit.com, "Sellnow G. Flashback."

17. Nobelprize.org, "Philip S. Hench: The Nobel Prize in Physiology or Medicine 1950," Nobel Foundation, 1950, http://nobelprize.org/nobel_prizes/ medicine/laureates/1950/hench-bio.html.

18. Philip S. Hench Walter Reed Yellow Fever Collection, "Letter from Philip Showalter Hench to Mrs. Forbes, July 2, 1942" (Charlottesville: Rector and Visitors of the University of Virginia, 1998–2004), http://etext.lib.virginia .edu/etcbin/ot2www-fever?specfile=/web/data/newreed/reed.02w&ac.

19. Hench would have preferred to be in Cuba studying yellow fever, but did what he could to participate in the war. In contrast, Ernest Hemingway *was*

in Cuba, but he seemed to be doing whatever he could to *avoid* participating in the war. He had already been involved in four of them (World War I, a small war between Greece and Turkey, the Spanish Civil War, and the Japanese-Chinese War, which he had covered as a correspondent along with his wife, Martha, shortly after their marriage). With his new facility at Finca Vigía, Hemingway was in a perfect position to drink, write, and sit this one out. But temptation reared its head, and Hemingway was never one to resist temptation too strenuously.

In 1943 there were reports of German U-boats patrolling the Gulf of Mexico in hopes of disrupting American shipping. The Hemingway Cookbook, "Dupuis K. Homing to the Stream: Ernest Hemingway in Cuba," DEE WHY BOOKS, 1999–2002; © Kelley Dupuis, http://www.deewhybooks .com.au/Magazine3/ReadingList.htm. Now that his home was in Cuba, it seemed the war had come to Hemingway. With the tacit endorsement of the American ambassador to Cuba, Hemingway outfitted his fishing boat, the *Pilar*, with machine guns and explosives; he put together a crew and began making excursions into the Gulf searching for German submarines. They never found one, but the hunt provided plenty of material for future Hemingway stories.

20. D. J. Ingle, *Edward C. Kendall, 1886–1972: A Biographical Memoir* (Washington, D.C.: National Academy of Sciences of the United States, 1975), 273.

21. E. C. Kendall, *Cortisone: Memoirs of a Hormone Hunter* (New York: Charles Scribner's Sons, 1971), 103.

CHAPTER 16. GOOD-BYE MARKER, HELLO SARETT

1. D. P. Steensma, "Luis Walter Alvarez: Another 'Mayo-Trained' Nobel Laureate," *Mayo Clinic Proceedings* 81, no. 2 (February 2006): 241–44, http:// www.mayoclinicproceedings.com/content/81/2/241.full.pdf+html.

2. M. Redig, "Yams of Fortune: The (Uncontrolled) Birth of Oral Contraceptives," *Journal of Young Investigators* 6 (2008), http://jyi.org/features/ ft.php?id=540.

3. L. Raber, "Steroid Industry Honored: International Historic Chemical Landmark Acclaims Success of Mexican Steroid Industry and a U.S. Chemist Who Made It Possible," *Journal of the Mexican Chemical Society* 43, no. 6 (1999): 235–37, http://redalyc.uaemex.mx/redalyc/html/475/47543610/47543610 .html.

4. G. S. Cohen, "Mexico's Pill Pioneer," *Perspectives in Health Magazine* 7, no. 1 (2002): 1–7, http://www.paho.org/english/dpi/Number13_article4_7.htm.

5. ACS Chemistry for Life, "The Marker Degradation" (Washington, D.C.: ACS Chemistry for Life), http://portal.acs.org/portal/acs/corg/content?_nfpb =true&_pageLabel=PP_ARTICLEMAIN&node_id=926&content _id=CTP_004452&use_sec=true&sec_url_var=region1&_uuid=.

6. American Chemical Society, "A Steroid Industry in Mexico" (Washington, D.C.: American Chemical Society, 2004), http://acswebcontent.acs.org/land marks/marker/mexico.html.

7. ACS Chemistry for Life, "Marker Degradation."

8. "Mexican Yams—Cortisone, Steroids, and Oral Contraceptives," http://www .botgard.ucla.edu/html/botanytextbooks/economicbotany/Dioscoreamed/ index.

9. Despite eventually receiving honors from the Mexican Chemical Society (1969) and the Chemical Congress of North America (1975), Russell Marker remained as unpredictable in his retirement as he was when he worked as a chemist. For many years the eccentric chemist disappeared from sight. Sporadic sightings occurred: he was often reported to be "in an insane asylum" or even dead. American Chemical Society, "Steroid Industry in Mexico." In reality, Marker was dividing most of his time between Mexico City and State College, Pennsylvania. He eventually became interested in objects made of silver. He spent time visiting various museums or collections by the great silversmiths, and commissioned Mexican reproductions of selected works. He also became philanthropic with the Pennsylvania State University, where he endowed an academic chair and several lectureships.

10. A. A. Patchett, "Lewis Hastings Sarett: December 22, 1917–November 29, 1999," *National Academy of Sciences, Biographical Memoirs*, vol. 81 (Washington, D.C.: National Academy Press, 2002), http://www.nap.edu/biomems/ lsarett.pdf; R. Hirschmann, "The Cortisone Era: Aspects of Its Impact. Some Contributions of the Merck Laboratories," *Steroids* 57, no. 12 (December 1992): 579–92, http://www.ncbi.nlm.nih.gov/pubmed/1481224.

11. Russell Marker left the University of Maryland without a PhD degree because he refused to take an extra course. Ironically, Lew Sarett was encouraged to skip a few courses and graduate early—he was given his PhD degree by Princeton after only two-and-a-half years because of the perceived urgent need to get him working on the large-scale production of steroids. Sarett's cutting-edge research at Princeton was simply too important to curtail merely to satisfy the formalities of a degree program; Marker would surely have been jealous.

12. E. C. Kendall, *Cortisone: Memoirs of a Hormone Hunter* (New York: Charles Scribner's Sons, 1971), 116–17.

13. L. H. Sarett, "Perhaps This Job Isn't Right for You," *R & D Innovator* 4, no. 5 (May 1995): 1–3, http://www.winstonbrill.com/bri1001/html/article_index/articles/151–200/article155_body.html.

CHAPTER 17. HENCH RETURNS TO MAYO

1. Nobelprize.org, "Philip S. Hench—Biography," Nobel Web AB, 2010, http://nobelprize.org/nobel_prizes/medicine/laureates/1950/hench-bio.html.
2. Philip S. Hench Walter Reed Yellow Fever Collection, "Letter from Philip Showalter Hench to Harry Schuman, January 26, 1946" (Charlottesville: Rector and Visitors of the University of Virginia, 1998–2004), http://yellowfever.lib.virginia.edu/reed/story.html.
3. Nowadays, unpublished authors have virtually no standing in the publishing community. Yet amazingly Hench was able to conclude his letter to the potential publisher with "Therefore, I cannot promise when I can finish the Walter Reed story. I can't even say whether it will be two years or five years, and since it is also indefinite, I simply do not want to commit myself at this time." Philip S. Hench Walter Reed Yellow Fever Collection, "Letter from Philip Showalter Hench to Harry Schuman, January 26, 1946."
4. Nobelprize.org, "Philip S. Hench. The Reversibility of Certain Rheumatic and NonRheumatic Conditions by the Use of Cortisone or of the Pituitary Adrenocorticotropic Hormone," Nobel Lecture, December 11, 1950, http://nobelprize.org/nobel_prizes/medicine/laureates/1950/hench-lecture.html.
5. P. Hanssen, "The Effect of Lactophenin-Icterus on Chronic Infectious Arthritis," *Acta Medica Scandinavica* 109, nos. 5–6 (1942): 494–506, http://www3.interscience.wiley.com/journal/122298673/abstract?CRETRY=1&SRETRY=0.
6. Sideburns, named in a roundabout fashion after Civil War general Ambrose Burnside, technically refer to the hair on the side of a man's face. However, Burnside and others typically allowed their "sideburns" to grow in continuity with an overgrown moustache. The chin was kept clean-shaven. The effect left observers wondering if the wearer had a squirrel tail wrapped across his upper lip.
7. faqs.org, "Carlos Juan Finlay Biography (1833–1915)," Advameg, 2010, http://www.faqs.org/health/bios/23/Carlos-Juan-Finlay.html.
8. Philip S. Hench Walter Reed Yellow Fever Collection, "Carlos Juan Finlay (1833–1915)" (Charlottesville: Rector and Visitors of the University of Virginia, 1998–2001), http://yellowfever.lib.virginia.edu/reed/finlay.html.

9. Philip S. Hench Walter Reed Yellow Fever Collection, "Letter from Philip Showalter Hench to Foster Kennedy, April 16, 1946" (Charlottesville: Rector and Visitors of the University of Virginia, 1998–2004), http://etext.lib .virginia.edu/etcbin/fever-browse?id=04106003.

10. R. Dallek, "The Medical Ordeals of JFK," *Atlantic Monthly*, December 2002, 49–61, http://www.theatlantic.com/past/issues/2002/12/dallek.htm.

11. Ibid.

12. Ibid.

13. Ibid.

14. Ibid.

15. Philip S. Hench Walter Reed Yellow Fever Collection, "Letter from Philip Showalter Hench to Foster Kennedy, April 16, 1946."

16. Philip S. Hench Walter Reed Yellow Fever Collection, "Letter from Philip Showalter Hench to Foster Kennedy, August 8, 1946" (Charlottesville: Rector and Visitors of the University of Virginia, 1998–2004), http://etext.lib .virginia.edu/etcbin/fever-browse?id=04112003.

17. Philip S. Hench Walter Reed Yellow Fever Collection, "Letter from Philip Showalter Hench to J. F. Siler, August 23, 1946" (Charlottesville: Rector and Visitors of the University of Virginia, 1998–2004), http://etext.lib.virginia .edu/etcbin/fever-browse?id=04112012.

18. Mayo Clinic Medical Library, December 19, 1947, "Doctor P. S. Hench: Please return the following overdue books immediately to the library: Southern Medical Journal, v. 32, 1939, Due Dec. 17, 1947," Cortisone Papers Collection, Mayo Clinic Historical Archives.

19. Philip S. Hench Walter Reed Yellow Fever Collection, "Letter from Philip Showalter Hench to P. I. Nixon, October 27, 1947" (Charlottesville: Rector and Visitors of the University of Virginia, 1998–2004), http://etext.lib .virginia.edu/etcbin/fever-browse?id=04136008.

20. This particular friend, however, did not find Hench to be without his faults. "He was often absentminded, as befits an intellectual mind in constant motion." But that was not the most poignant observation. "There was only one circumstance in which a Frenchman found it difficult to please him—and that was meal time. He set up for himself . . . a particular diet from which he would not depart no matter what country he traveled in; raw vegetables, fresh fruit, cold cuts, smoked fish, etc. and as a beverage, iced tea. It was useless to invite him to a French meal; at the table he thanked the hostess very graciously but avoided eating the various courses which were offered to him." The Frenchman went on to describe a particularly vexing episode in which he took Dr. Hench to lunch at a well-known restaurant in Paris's Latin Quarter. After much cajoling the rheumatologist agreed to eat a bean casserole—"*cassoulet*

toulousain"—but instead of drinking it with wine, as everyone else in the world would have done, Hench had his with a large glass of cold milk. J. Forrestier, "Philip Hench," *Médecine et hygiène* 23 (1965): 435.

21. Ibid.
22. D. J. Ingle, *Edward C. Kendall, 1886–1972: A Biographical Memoir* (Washington, D.C.: National Academy of Sciences of the United States, 1975), 268.
23. Ibid.
24. H. Butt, oral communication, January 2007.
25. Ingle, *Edward C. Kendall*, 268.

CHAPTER 18. PUSH ON? GIVE UP?

1. D. J. Ingle, *Edward C. Kendall, 1886–1972: A Biographical Memoir* (Washington, D.C.: National Academy of Sciences of the United States, 1975).
2. E. C. Kendall, *Cortisone: Memoirs of a Hormone Hunter* (New York: Charles Scribner's Sons, 1971), 114–15.
3. Ibid.
4. Ibid. Taken in historical context, this remark seems incredible—it was during the March meeting in Atlantic City, just hours before the disastrous scientific session dealing with compound A took place, that Kendall received word his son Roy had just died of leukemia. Most people would have dropped what they were doing and rushed home. Not Kendall. He seemed more concerned with his chemistry crisis. Was this just his way of coping with intense loss and personal pain? Or was Kendall so dedicated to his work that everything else in his life—including family—came in a distant second?
5. Ibid.
6. R. Hirschmann, "The Cortisone Era: Aspects of Its Impact. Some Contributions of the Merck Laboratories," *Steroids* 57, no. 12 (December 1992): 579–92, http://www.ncbi.nlm.nih.gov/pubmed/1481224.
7. L. H. Sarett, "Partial Synthesis of Pregnene-4-TRIOL-17(β), 20(β), 21-DIONE-3, 11 and Pregnene-4-DIOL-17(β),21-TRIONE-3,11,20 Monoacetate," *Journal of Biological Chemistry* 162 (1946): 601–32, http://www.jbc.org/.
8. The Merck Index: Kendall-Mattox Reaction (Whitehouse Station, N.J.: Merck Sharp & Dohme, 2010), http://themerckindex.chemfinder.com/TheMerckIndex/NameReactions/displayonr.asp?onr=ONR211.htm.
9. Kendall, *Cortisone*, 117.

10. S. G. Hillier, "Diamonds Are Forever: The Cortisone Legacy," *Journal of Endocrinology* 195 (2007): 1–6, http://joe.endocrinology-journals.org/cgi/content/full/195/1/1.

11. After nearly two decades of search for *the* cortin, investigators were still empty-handed. Why? Because there was no such thing as *the* cortin. It was becoming clear that the adrenal cortex made a number of steroid substances, none of which were sufficient by themselves to maintain life when the adrenals were removed. At least two distinct types of adrenal product were necessary. One type, similar to DOCA (the compound John Kennedy may have been taking), was called a mineral corticoid—these steroids affect blood pressure, along with the excretion of salt (a mineral) and water. The other group, best exemplified by compound E, included "glucocorticoids," which affect sugar (glucose) and fat metabolism, the contractile strength of muscle, and many other body functions. This explained why extracts of the adrenal cortex were so effective at treating Addison's disease—extracts contained both types of steroid.

12. H. F. Polley and C. H. Slocumb, "Cortisone—A Historical Discovery," Mayo Historical Records, Mayo clinic Archives, Plummer Building.

13. Kendall, *Cortisone*, 119–20.

14. Ibid.

15. Ibid.

16. Ibid.

17. Ibid.

CHAPTER 19. THE DECISION TO TEST COMPOUND E ON RHEUMATOID ARTHRITIS

1. H. F. Polley, C. H. Slocumb, Cortisone—a historical discovery, unpublished notes, Mayo Clinic Archives.

2. P. S. Hench, "Potential Reversibility of Rheumatoid Arthritis," *Annals of the Rheumatic Diseases* 8, no. 2 (June 1949): 90–96.

3. E. C. Kendall, *Cortisone: Memoirs of a Hormone Hunter* (New York: Charles Scribner's Sons, 1971), 122, 124–25; H. F. Polley and C. H. Slocumb, "Behind the Scenes with Cortisone and ACTH," *Mayo Clinic Proceedings* 51, no. 8 (August 1976): 471–77; Polley and Slocumb, Cortisone—a historical discovery; H. F. Polley, Cortisone [lecture], unpublished Mayo Clinic Archives, 1975; H. F. Polley, Cortisone [rough draft], staff memoirs, unpublished Mayo Clinic Archives, 1975[0][0].

4. Kendall, *Cortisone*, 122, 124–25.

5. Ibid.
6. P. S. Hench, Letter from Philip Showalter Hench to Dr. Augustus Gibson, September 4, 1948, Mayo Clinic Archives.
7. Polley and Slocumb, Cortisone—a historical discovery.
8. Literature.org: The Online Literature Library, *The Sign of Four,* by Arthur Conan Doyle, http://www.literature.org/authors/doyle-arthur-conan/sign -of-four/.
9. Once again, Kendall was wrong—Dr. Slocumb had been giving doses of 100 milligrams a day to his patients.
10. Polley and Slocumb, Cortisone—a historical discovery.
11. Kendall, *Cortisone,* 122, 124–25.
12. Polley and Slocumb, Cortisone—a historical discovery.

CHAPTER 20. THE AMAZING MRS. G.

1. H. F. Polley and C. H. Slocumb, "Behind the Scenes with Cortisone and ACTH," *Mayo Clinic Proceedings* 51, no. 8 (August 1976): 471–77; E. C. Kendall, *Cortisone: Memoirs of a Hormone Hunter* (New York: Charles Scribner's Sons, 1971), 125–26.
2. William Heberden (1710–1801) was a London physician remembered for, among other things, his interest in rheumatological disorders.
3. Kendall, *Cortisone,* 125–26.
4. H. F. Polley, Cortisone [rough draft], 1975, staff memoirs, unpublished, Mayo Clinic Archives.
5. Ibid.
6. H. F. Polley, Cortisone [rough draft].
7. H. F. Polley, Cortisone [lecture], 1975, unpublished, Mayo Clinic Archives.
8. Ibid.
9. Polley, Cortisone [rough draft].
10. Kendall, *Cortisone,* 125–26.
11. Ibid.
12. Polley, Cortisone [lecture].
13. Polley, Cortisone [rough draft].

CHAPTER 21. A PROMISING START

This chapter is based largely on the unpublished notes of H. F. Polley and C. H. Slocumb from the Mayo Clinic Archives.

1. E. C. Kendall, *Cortisone: Memoirs of a Hormone Hunter* (New York: Charles Scribner's Sons, 1971), 126–27, 132, 134–35.
2. H. F. Polley and C. H. Slocumb, "Behind the Scenes with Cortisone and ACTH," *Mayo Clinic Proceedings* 51, no. 8 (August 1976): 471–77.
3. Ibid.
4. Ibid.
5. Ibid.
6. Kendall, *Cortisone*, 126–27, 132, 134–35; Polley and Slocumb, "Behind the Scenes," 471–77.
7. Kendall, *Cortisone*, 126–27, 132, 134–35.
8. Ibid.
9. Ibid.

CHAPTER 22. THE BAD AND THE UGLY

This chapter is based largely on the unpublished notes of H. F. Polley and C. H. Slocumb in the Mayo Clinic Archives.

1. They were successful. Almost a year later, when work with compound E expanded to involve other physicians in other locations, an opportunity arose for the Mayo team to discuss the situation with their Chicago colleague. He knew nothing about the events at Mayo, and "it was only then that we knew that she had, indeed, honored our request."
2. H. F. Polley and C. H. Slocumb, "Behind the Scenes with Cortisone and ACTH," *Mayo Clinic Proceedings* 51, no. 8 (August 1976): 471–77.
3. Although the development of a "lupuslike syndrome" is known today as a potential side effect of compound E treatment, at that time it simply looked like she had been misdiagnosed at the Mayo Clinic as having rheumatoid arthritis when she really had lupus.
4. E. L. Matteson and J. L. Brown, "A Patient Remembers the Miracle Drug Cortisone 50 Years Later," *Minnesota Medicine* 85, no. 6 (June 2002): 12–15.
5. As a result of the gold injections, Brown recalled yellow tears running from her eyes.

6. Despite her devastating disease, Brown kept a good attitude about life and remained reasonably independent. She married (her husband had developed polio as a child) and eventually conceded that "all things considered, we've had a relatively good life."

7. S. Wasson, "Bigger Than Life: The Picture, the Production, the Press," *Senses of Cinema* 58 (1999–2009), http://archive.sensesofcinema.com/contents/06/38/bigger_than_life.html.

8. Compound E wasn't the only toxic material here; the movie itself was box-office poison. *Bigger Than Life* has disappeared completely from the public eye.

CHAPTER 23. PROGRESS AND SETBACKS

1. H. F. Polley and C. H. Slocumb, Cortisone—a historical discovery, unpublished notes, Mayo Clinic Archives.

2. Ibid.

3. E. C. Kendall, *Cortisone: Memoirs of a Hormone Hunter* (New York: Charles Scribner's Sons, 1971), 136–37.

4. H. F. Polley and C. H. Slocumb, "Behind the Scenes with Cortisone and ACTH," *Mayo Clinic Proceedings* 51, no. 8 (August 1976): 471–77.

5. Ibid.

6. Kendall, *Cortisone*, 136–37.

7. Unfortunately, the side effects of ACTH seemed more troublesome than those associated with compound E. Fluid retention (like that seen with Mrs. G.'s fatal pulmonary edema), skin discoloration, and other undesirable effects tended to be more common and more severe with ACTH. In retrospect, this makes perfect sense; ACTH causes the adrenal gland to release many substances in addition to compound E, most of which are not beneficial for the condition under treatment. ACTH is also difficult to purify, and therefore the injected material was frequently contaminated with unwanted animal proteins, some of which triggered their own unique side effects. Ultimately, the trials with ACTH further confirmed the rationale for using compound E; ACTH, however, offered physicians an alternative to the scarce, expensive adrenal steroid.

8. Kendall, *Cortisone*, 136–37.

9. Ibid.

10. Polley and Slocumb, "Behind the Scenes," 471–77.

11. Ibid.

12. H. F. Polley, Cortisone [5], 1975, staff memoirs, unpublished, Mayo Clinic Archives.
13. Ibid.
14. Ibid.
15. Ibid.
16. M. E. Warner, "Witness to a Miracle: The Initial Cortisone Trial: An Interview with Richard Freyberg, MD," *Mayo Clinic Proceedings* 76 (2001): 529–32.
17. Ibid., 531–32.
18. Polley, Cortisone [6].
19. H. F. Polley, Cortisone [rough draft], 1975, staff memoirs, unpublished, Mayo Clinic Archives.
20. Polley, Cortisone [13–14].

CHAPTER 24. CONVINCING THE SKEPTICS

1. H. F. Polley, Cortisone [rough draft], [12–13], 1975, staff memoirs, unpublished, Mayo Clinic Archives.
2. M. E. Warner, "Witness to a Miracle: The Initial Cortisone Trial: An Interview with Richard Freyberg, MD," *Mayo Clinic Proceedings* 76 (2001): 529–32.
3. Ibid.
4. Ibid.
5. Ibid.
6. Ibid.
7. Ibid. H. F. Polley, Cortisone [9], 1975, staff memoirs, unpublished, Mayo Clinic Archives. Notwithstanding the occasional friendly conflict, "it was a week of miraculous discovery, exchange of information, and collegiality."
8. The results from these trials turned out even better than anyone could have hoped; every patient in each of the five geographically separate practices experienced dramatic improvement following compound E administration. The ability of compound E to put rheumatic disease into remission was no longer in doubt.
9. Polley, Cortisone [9].
10. Warner, "Witness to a Miracle," 529–32.
11. H. Butt, oral communication, January 12, 2007.

CHAPTER 25. ANNOUNCEMENT

This chapter is based largely on the unpublished notes of H. F. Polley and C. H. Slocumb in the Mayo Clinic Archives.

1. H. Butt, oral communication, January 12, 2007.
2. Given the lack of effective medical or surgical treatment for rheumatoid disease, physical therapy was the main treatment modality for these patients. It was an area in which every graduating rheumatology fellow needed expertise.
3. Kemper denied any specific knowledge of the secret trials at Mayo. He had probably heard loose "talk" around the clinic and may have even picked up some general, highly speculative information during conversations two weeks earlier at a medical meeting in Boston, but he certainly did not know that compound E was the therapy being tested.
4. Just a few years earlier the reporter had shared a unique honor with Luis Alvarez—he'd been allowed to witness the first A-bomb explosion in New Mexico. Alvarez may have been one of the few physicists to see the first mushroom cloud, but Laurence was the only member of the Fourth Estate allowed to do so. Why? Because Laurence represented the powerful *New York Times*, and perhaps more important, because his journalistic reputation was "above reproach."
5. E. C. Kendall, *Cortisone: Memoirs of a Hormone Hunter* (New York: Charles Scribner's Sons, 1971), 141.
6. Ibid.
7. Butt, oral communication.
8. Kendall, *Cortisone*, 142.
9. The Lasker Foundation—Awards, "Albert Lasker Medical Journalism Awards: 1949; William L. Laurence," Lasker Foundation, 2009, http://www.lasker foundation.org/awards/formaward.htm.
10. Kendall, *Cortisone*, 144.
11. Ibid.
12. Ibid.
13. Kendall clearly interpreted the results of his approach as success, noting: "In a short time conflict and confusion were replaced with good will and harmony."
14. Kendall, *Cortisone*, 145.
15. Ibid.
16. Ibid.
17. Ibid.

CHAPTER 26. THE PRIZE

1. D. P. Steensma, "Luis Walter Alvarez: Another 'Mayo-Trained' Nobel Laureate," *Mayo Clinic Proceedings* 81, no. 2 (February 2006): 241–44, http://www.mayoclinicproceedings.com/content/81/2/241.full.pdf+html.

2. Nobelprize.org, "Luis Alvarez: The Nobel Prize in Physics 1968 (Biography)," The Nobel Foundation, 1968, http://nobelprize.org/nobel_prizes/physics/laureates/1968/alvarez-bio.html.

3. And perhaps the soon-to-be-popular Compound W® wart remover?

4. E. C. Kendall, *Cortisone: Memoirs of a Hormone Hunter* (New York: Charles Scribner's Sons, 1971), 143, 151–52.

5. D. Wyre and J. R. Williams, "Percy Lavon Julian—Life and Legacy," http://www.temple.edu/chemistry/main/faculty/Williams/Percy%20Lavon%20Julian%201.htm.

6. Julian deserved to be hailed as a hero for his accomplishments. But that wasn't always the case. While working through these exciting new steroid synthetic pathways, he and his family moved to the Chicago suburb of Oak Park, Illinois—the birthplace and hometown of Ernest Hemingway. Julian quickly discovered that blacks were not welcomed here. Racism was rampant in the largely stale, pale, and run by males community. Even before the family moved in, their home was fire bombed. A year later the house was damaged by an explosion when ill-wishers detonated dynamite in their yard. Community support eventually formed around the Julians, but it was well known that Percy and his son often felt it necessary to stand sentry duty around their home—with shotguns. B. Witkop, "Percy Lavon Julian: April 11, 1899–April 19, 1975" (Washington, D.C.: National Academies Press, Biographical Memoirs), http://www.nap.edu/html/biomems/pjulian.html/; Science Alive, For Teachers: "The Life and Science of Percy Julian" (S. Lyons, "The Producer's Story: Rediscovering a Forgotten Genius," WGBH Educational Foundation, 2007), http://www.pbs.org/wgbh/nova/julian/producer.html; Philadelphia: Chemical Heritage Foundation, 2005, http://www.chemheritage.org/scialive/julian/teachers/narrative.html.

7. The Story of the Mexican Wild Yam (Dioscorea Barbasco): A Natural Source of Progesterone, "Who Discovered Progesterone Could Be Made from a Plant Source?" Progesterone Research Network, 2006–2010, http://www.progesteroneresearchnetwork.com/HTML/articles/files/060505-who-discovered-progesterone-could-be-made-from-a-plant-source.html/.

8. Shunya's Notes, "Percy Julian, Chemist Extraordinaire," June 24, 2007, http://blog.shunya.net/shunyas_blog/2007/06/percy-julian-ch.html.

9. Ibid.

10. "DePauw University Celebrates Percy Lavon Julian '20: Chemist, Scholar, Entrepreneur, Civic Leader, Humanitarian, 1899–1975" (Newcastle, Ind.: DePauw University, 2000–2010), http://www.depauw.edu/library/archives/percyjulian/.
11. Shunya's Notes, "Percy Julian."
12. Ibid.
13. Ibid.
14. Archives of DePauw University and Indiana United Methodism, Percy Lavon Julian, 1899–1975: DePauw University Class of 1920, Percy Lavon Julian '20 Family Papers 1899–1975, Biographical Sketch (Newcastle, Ind.: DePauw University, June 2007), http://www.depauw.edu/library/archives/dpuinventories/julian_percy_lavon_family.htm.
15. Kendall, *Cortisone*, 143, 151–52.
16. H. F. Polley and C. H. Slocumb, Cortisone—a historical discovery, unpublished notes [18], Mayo Clinic Archives.

CHAPTER 27. STOCKHOLM

Much of this chapter is from Dr. Kendall's book (chapter 14), unpublished archive notes, and Dr. Hench's memoir publication no. 2.

1. E. C. Kendall, *Cortisone: Memoirs of a Hormone Hunter* (New York: Charles Scribner's Sons, 1971), chapter 14.
2. Nobelprize.org, "Reminiscences of the Nobel Festival, 1950, by Philip S. Hench," November 6, 2001, http://nobelprize.org/award_ceremonies/ceremony _sthlm/eyewitness/hench/index.html.
3. Ibid.
4. Ibid.
5. Ibid.
6. Kendall, *Cortisone*, chapter 14.
7. Nobelprize.org, "Reminiscences of the Nobel Festival."
8. Nobelprize.org, "The Nobel Prize in Physiology or Medicine 1950: Presentation Speech by Professor G. Liljestrand," Nobel Foundation, 1950, http://nobelprize.org/nobel_prizes/medicine/laureates/1950/press.html.
9. Nobelprize.org, "Reminiscences of the Nobel Festival."
10. Nobelprize.org, "Edward C. Kendall: The Nobel Prize in Physiology or Medicine 1950," Banquet Speech, Nobel Foundation, 1950, http://nobelprize .org/nobel_prizes/medicine/laureates/1950/kendall-speech.html.
11. Ibid.

12. Ibid.

13. H. Butt, oral communication, January 2007.

CHAPTER 28. AFTERMATH

1. Edmund Fitzgerald, Insurance Executive, *New York Times*, Obituary, Late City Final Edition, January 10, 1986, http://select.nytimes.com/gst/abstract .html?res=F50715FF355F0C738DDDA80894DE484 . . . /.

2. "Ship Named in Honor of Mayo Association Member," *Mayovox* 9, no. 17 (1958).

3. "Officers Named by Association, Legal Counsel—Harry A. Blackmun," *Mayovox* 2, no. 6 (February 3, 1951).

4. Philip S. Hench Walter Reed Yellow Fever Collection, "Remarks Introducing Philip Showalter Hench to the Rotary Club [of Havana], January 1952" (Charlottesville: Rector and Visitors of the University of Virginia, 1998–2001), http://yellowfever.lib.virginia.edu/reed/story.html.

5. Philip S. Hench Walter Reed Yellow Fever Collection, "Interview with Philip Showalter Hench by a Cuban Newspaper, [1952]: Answers to Newspaper Questions, Dr. Philip S. Hench, the Mayo Clinic, Rochester, Minnesota" (Charlottesville: Rector and Visitors of the University of Virginia, 1998–2001), http://yellowfever.lib.virginia.edu/reed/story.html.

6. Ibid.

7. S. A. Paget, M. Lockshin, and S. Loebl, *The Hospital for Special Surgery Rheumatoid Arthritis Handbook* (Hoboken, N.J.: John Wiley and Sons, 2001), 98, http://books.google.com/books?id=akacOyQrnKkC&source=gbs _navlinks_s/.

8. P. S. Hench, E. C. Kendall, C. H. Slocumb, and H. F. Polley, "The Effect of a Hormone of the Adrenal Cortex (17-Hydroxy-11-Dehydrocorticosterone; Compound E) and of Pituitary Adrenocorticotropic Hormone on Rheumatoid Arthritis," *Mayo Clinic Proceedings* 24, no. 8 (April 13, 1949): 181–97.

9. Philip S. Hench Walter Reed Yellow Fever Collection, "Interview with Philip Showalter Hench by a Cuban Newspaper."

10. H. F. Polley, Cortisone, staff memoirs, 1975, unpublished, Mayo Clinic Archives.

11. J. Hench, Letter to T. W. Rooke, April 30, 2007.

12. Ibid.

13. Ibid.

14. Ibid.

15. D. J. Ingle, *Edward C. Kendall, 1886–1972: A Biographical Memoir* (Washington, D.C.: National Academy of Sciences of the United States, 1975).

16. Nobelprize.org, "Ernest Hemingway: The Noble Prize in Literature 1954: Banquet Speech, December 10, 1954," Nobel Media AB, 2010, http://nobelprize.org/nobel_prizes/literature/laureates/1954/hemingway-speech.html.

17. L. F. Bunting, "The Short Unhappy Life of Ernest Hemingway," *Counterweights Magazine*, March 19, 2009, http://www.counterweights.ca/2009/03/hemingway/.

18. R. Dallek, "The Medical Ordeals of JFK," *Atlantic Magazine*, December 2002, http://www.theatlantic.com/magazine/archive/2002/12/the-medical-ordeals-of-jfk/5572/4/.

19. Ibid.

20. In 1955 an article appeared in the American Medical Association's archives of surgery entitled "Management of Adrenal-Cortical Insufficiency during Surgery." It described a thirty-seven-year-old man who was operated on for "serious back pain" and became the first patient with Addison's disease to survive a surgery of this magnitude. In 1961 it was publicly revealed that the patient in question was indeed the then-current president, John F. Kennedy.

21. Dallek, "The Medical Ordeals of JFK."

22. "The Straight Dope: Did John F. Kennedy Really Write 'Profiles in Courage'?" November 7, 2003, Creative Loafing Media, 1996–2010, http://www.straightdope.com/columns/read/2478/did-john-f-kennedy-really-write-profiles-in-courage/.

CHAPTER 29. TWILIGHT

1. E. Ward, interview by author, October 22, 2007.

2. "Memorandum from Philip Showalter Hench, February 24, 1954," University of Virginia Collection of Hench Letters (Charlottesville: Rector and Visitors of the University of Virginia, 1998–2004).

3. Ibid.

4. H. F. Polley, Cortisone, staff memoirs, 1975, unpublished, Mayo Clinic Archives.

5. Council on Pharmacy and Chemistry, "Report of the Council: Annual Meeting of the Council on Pharmacy and Chemistry," *Journal of the American Medical Association* 142, no. 5 (February 4, 1950): 339–40.

6. One publication reported that at least one patient had committed suicide because of the effects of cortisone.

7. M. C. Borman and H. C. Schmallenberg, "Suicide Following Cortisone Treatment," *Journal of the American Medical Association* 146 (1951): 33738.

8. "Effects of Cortisone and Pituitary Adrenal Corticotropic Hormone" (editorial), *Journal of the American Medical Association* 142, no. 10 (March 11, 1950): 730–31.

9. "Hormones in Arthritis" (editorial), *New England Journal of Medicine* 243 (1950): 166–67.

10. R. H. Freyberg, C. H. Traeger, M. Patterson, W. Squires, C. H. Adams, and C. Stevenson, "Problems of Prolonged Cortisone Treatment for Rheumatoid Arthritis: Further Investigations," *Journal of the American Medical Association* 147, no. 16 (December 15, 1951): 1538–43.

11. N. Gutteridge, "American Medical Association Assembly" (special article), *Lancet* 258, no. 6671 (July 1951): 31–33.

12. "A Comparison of Cortisone and Aspirin in the Treatment of Early Cases of Rheumatoid Arthritis: A Report by the Joint Committee of the Medical Research Council and Nuffield Foundation on Clinical Trials of Cortisone, A.C.T.H., and Other Therapeutic Measures in Chronic Rheumatic Diseases," *British Medical Journal* 1, no. 4873 (May 29, 1954): 1223–27.

13. J. H. Glyn, "The Discovery of Cortisone: A Personal Memory," *BMJ* 317, no. 7161 (September 19, 1998): 822A

14. Ward, interview.

15. How close? Close enough that Hench was asked to help name the Glyns' newborn daughter. She was eventually given a female version of "Philip," but only after the name "Cortisona" was suggested by Hench—and rejected by the parents. Glyn, "The Discovery of Cortisone," 822A.

16. J. Glyn, "The Discovery and Early Use of Cortisone," *Journal of the Royal Society of Medicine* 91 (October 1998): 513–17.

17. Glyn, "Discovery of Cortisone," 822A.

18. J. Hench, Letter to T. W. Rooke, April 30, 2007.

19. The society was so named for two reasons: because Holmes had assumed the alias of a Norwegian explorer while hiding out after his purported death at Reichenbach Falls, and Norwegians comprise the predominant ethnicity in Minnesota.

20. "Two City Men Contribute to Book on Sherlock Holmes," *Rochester PostBulletin,* January 30, 1958.

21. It's been speculated that the FBI had been watching Hemingway since his submarine hunting escapades during the war, and that ongoing surveillance to assess his relationship with Castro had been in place since then. K. Dupuis, *Homing to the Stream: Ernest Hemingway in Cuba*, Caroline Hulse, 1999–2006, http://www.ernest.hemingway.com/cuba.htm.

22. C. Baker, ed., *Ernest Hemingway: Selected Letters, 1917–1961* (New York: Scribner, 2003), 916.

23. The events transpiring between Ernest Hemingway, the Mayo Clinic, the doctors who treated him in Rochester, the effects of his antidepression therapy, the government's role in his hospitalization, and many other aspects of this tragic saga are debatable, disputable, and ultimately not resolvable. What appears most likely, based on publicly available information, is that Hemingway was hospitalized in Rochester through the beginning of 1961, and that during this time he received electroconvulsive therapy—perhaps as many as fifteen sessions. As a result of this treatment he improved enough to be dismissed from the hospital; no one ever claimed he was "cured" of his depression. Unfortunately, Hemingway faced continuing mental problems after his discharge, and he became convinced that the electroshock therapy had destroyed both his memory and his ability to write. If this was true, it was likely that any relief from his depression would be temporary.

24. Glyn, "Discovery and Early Use of Cortisone," 513–17.

25. R. W. Nickeson, "Philip Showalter Hench, Nobel Laureate," *Alumni News University of Pittsburgh Medical Center* (Spring 1993): 14–16.

26. J. Hench, Letter to T. W. Rooke, April 30, 2007.

27. E. H. Rynearson, Memo to J. Eckman, May 10, 1965, Mayo Clinic Historical Archives.

28. Glyn, "Discovery of Cortisone," 822A.

CHAPTER 30. THE END OF THE SHOW

1. Percy Lavon Julian—Citizendia, Citizendia.org, 2009, http://www.citizen dia.org/Percy_Lavon_Julian.

2. Ernest Hemingway: Nobel Prize, Lycos Inc, 2010, http://www.lycos.com/info/ernest-hemingway—nobel-prize.html; D. Smith, "Shock and Disbelief," *Atlantic Monthly*, February 2001, 1–9; B. Kert, *The Hemingway Women* (New York: W. W. Norton, 1983), 499; J. Meyers, *Hemingway: A Biography* (New York: Harper & Row, 1985), 550–52.

3. Kert, *Hemingway Women*, 499; Meyers, *Hemingway*, 550–52.

4. Jim Garrison, the flamboyant New Orleans attorney best known for his attempt to prosecute certain nefarious individuals for the assassination of President John Kennedy, may have raised yet another possibility: that there was a Hemingway plot involving J. Edgar Hoover, the FBI, and Hemingway's psychiatrist. Garrison's convoluted speculations seem to insinuate that

Hemingway's release from Saint Marys could have been influenced by a powerful, government-aligned cartel intent on harming the author.

5. "Hemingway Dead of Shotgun Wound: Wife Says He Was Cleaning Weapon," Special to the *New York Times*, July 3, 1961 (New York Times Company, 1999), http://www.nytimes.com/books/99/07/04/specials/hemingway-obit.html.

6. Hemingway claimed that the electroconvulsive therapy treatments he'd undergone had destroyed his memory along with his ability to write, and in doing so "put him out of business." Recognizing the significant role of mental illness in his suicide, the Roman Catholic Church deemed him "not mentally responsible" for his own demise and allowed the Nobel-winning author to be buried in a religious service. WordPress.com, "The Life of Hemingway," July 25, 2009, http://thelifeofhemingway.wordpress.com/.

7. R. Dallek, "The Medical Ordeals of JFK," *Atlantic Magazine,* December 2002, http://www.theatlantic.com/past/docs/issues/2002/12/dallek.htm.

8. L. W. Alvarez, *Alvarez: Adventures of a Physicist* (New York: Basic Books, 1987), 13–14.

9. "A Physicist Examines the Kennedy Assassination Film," *American Journal of Physics* 44, no. 9 (September 1976): 813–27, http://scitation.aip.org/vsearch/servlet/VerityServlet?KEY=AJPIAS&ONLINE=YES&smod/.

10. C. E. Yesalis and M. S. Bahrke, "History of Doping in Sport," in *Performance-Enhancing Substances in Sport and Exercise*, ed. M. S. Bahrke and C. E. Yesalis (Champaign, Ill.: Human Kinetics, 2002), 7.

11. Bill Toomey Official Web site—The Athlete, "Bill Toomey Puts the Decathlon into Perspective," William A. Toomey, 2001, http://www.billtoomey.com/Athlete.html.

12. The Nobel Prize in Physics 1968: Luis Alvarez, Nobelprize.org, http://nobelprize.org/nobel_prizes/physics/laureates/1968/.

13. Luis W. Alvarez (American Physicist), Encyclopedia Britannica Online, http://www.britannica.com/EBchecked/topic/18131/Luis-W-Alvarez/.

14. C. G. Wohl, "Scientist as Detective: Luis Alvarez and the Pyramid Burial Chambers, the JFK Assassination, and the End of the Dinosaurs," http://www.6911norfolk.com/d01bln/105f06/105f06-wohl-alvarez.pdf.

15. D. J. Ingle, *Edward C. Kendall, 1886–1972: A Biographical Memoir* (Washington, D.C.: National Academy of Sciences of the United States, 1975), 273.

16. Ibid.

17. But perhaps not as famous as his sister? On March 13, 1985, Mabel Alvarez, Walter's younger sister, was at peace with the world. She'd been in a Los Angeles nursing home for two years after a fall left her with a painful broken hip.

The walls of her room were covered with her favorite paintings. Art increases in value after the death of the artist, and unfortunately these paintings would soon be worth millions.

An old friend visited Mabel that final evening: as he read to her from a book by Robert Louis Stevenson, he suddenly paused and commented, "Mabel, I've just thought of something. In future times, Walter Alvarez will be remembered only as Mabel Alvarez' brother. His books have been out of print for years, but your work will continue to delight people for centuries." The ninety-four-year-old personification of timeless, dignified femininity looked up, and in a "clear, strong voice," said, "You know, I never thought of that!" She died in her sleep a few hours later. Mabel Alvarez Estate Collection— Glenn Bassett Essay, "Mabel Alvarez (1891–1985): A Personal Memory," http://www.mabelalvarez.com/about/bassett.htm.